Calculators are proving to be just as useful to the astronomer as they have been to other scientists, and this book shows how a modern electronic calculator can best be used for astronomical problems which demand accurate solutions in a minimum of time. Subjects covered include sidereal and mean time; reduction for precession; proper motion; reduction to apparent place; binary star orbits; coordinates of comets from parabolic and elliptical elements; Least Squares; equation of the equinoxes; and planetary coordinates. Methods are demonstrated for both algebraic logic calculators and those employing Reverse Polish Notation. The extensive appendix contains fifty-seven fully documented programmes for key- and magnetic card-programmable calculators. In addition to topics dealt with in the main text, these programmes cover: lunar eclipses and occultations; interpolation; Chebyshev polynomials; Julian Date; iteration for transcendental variables; coordinate conversions; and many others. The reader will be saved many hours of valuable time in writing, testing and de-bugging astronomical programmes.

Throughout the book, worked examples are given with full step-by-step guidance and all the necessary equations, definitions and reference sources. Further practice problems are set (with solutions).

No astronomer or student can afford to be without his own copy of this unique book.

Aubrey Jones is a Fellow of the Royal Astronomical Society and a member of the British Astronomical Association. His main interests are double stars, lunar occultations, planetary studies and astrophotography. In the fibreglass dome of his observatory at Rainham (Kent) he uses a 10in Newtonian reflector. As an accredited Webb Society observer at the Old Greenwich Observatory of the National Maritime Museum he measures double stars with the 28in refractor.

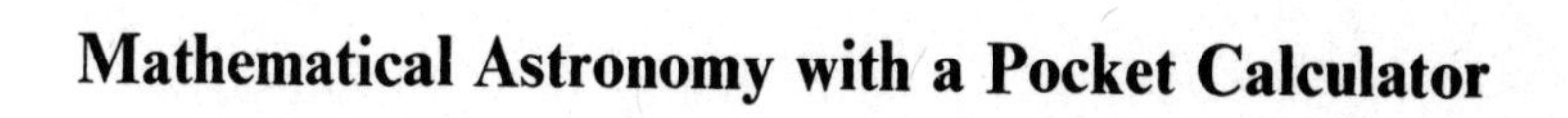

Mathematical Astronomy with a Pocket Calculator

A HALSTED PRESS BOOK

Mathematical Astronomy with a Pocket Calculator

Aubrey Jones FRAS

JOHN WILEY & SONS
New York

Published in the U.S.A.
by Halsted Press, a Division of
John Wiley & Sons, Inc., New York.

Library of Congress Cataloging in Publication Data
Jones, Aubrey.
Mathematical astronomy with a pocket calculator.
"A Halsted Press book."
1. Astronomy – Mathematics. 2. Calculating-machines.
I. Title.
QB47.J66 1978 520'.1'51 78–12075
ISBN 0–470–26552–3

Printed in Great Britain

Contents

Preface

For some modern astronomers, the mathematical aspect of astronomy is an anathema to be avoided wherever possible—because they find it dull and uninteresting, or because it is too time-consuming, or because it is difficult to avoid making mistakes or to trace and correct them when they have been made. But few, if any, can manage to escape altogether from the need for computation of one kind or another. To others, astronomy *is* mathematics, with comprehensive computer facilities equal in importance to the observational instrumentation. For the majority, somewhere between these two extremes, mathematical astronomy can be either a delight and a pleasure which can be indulged on cloudy nights, or something of a chore which must be accomplished in the minimum of time.

This book is intended for the working astronomer, and is shaped to meet his needs. It assumes, therefore, that the observer is fully aware of the nature of the problem in hand but seeks guidance in the application of a relatively new tool which he can employ to obtain an accurate solution to the problem, either at the eyepiece or in the preparation (or completion) of an observing programme.

It is clearly not the purpose of this book to supplant such recognized works as Chauvenet's *A Manual of Spherical and Practical Astronomy*, Newcomb's *A Compendium of Spherical Astronomy* and Smart's *Spherical Astronomy*, or (more recently) Woolard and Clemence's *Spherical Astronomy* and McNally's *Positional Astronomy*. Neither will it take the place of the *Astronomical Ephemeris* (*American Ephemeris and Nautical Almanac* in the USA) or its invaluable *Explanatory Supplement* (although it may, under certain conditions, free the observer from total dependence on such an ephemeris). For those who require more detailed treatments and explanations there will always be a place on the bookshelf for these works.

The current aim is to set out, in readily accessible form, methods of calculating such items as the Local Sidereal Time for any geographical location, allowance for the effects of precession on star positions and proper motions, ephemerides for visual binary stars and comets, and so on. Fully worked examples are given. Normally, each example is worked twice: first, there is a method to suit those who have the use of a straightforward scientific-type pocket electronic calculator, with algebraic logic and limited memory facilities; this is followed by a method which exploits to the full the facilities provided by more advanced types of pocket calculator, using Reverse Polish Notation (RPN) and multiple user-addressable memory stores.

The electronic-calculator styles used in this book are the basic algebraic and the Hewlett-Packard RPN keyboard systems. It is obviously impossible to give detailed instructions which will suit all the various keyboard layouts and memory facilities offered by the many different manufacturers. The two systems selected will cover the needs of the majority of readers, and others will be able to make occasional modifications should it not be possible to execute an instruction as printed: for instance, on the very cheapest of scientific calculators it might be found that trigonometrical functions can be performed only on angles of the first quadrant; with a little ingenuity the possessor of such a calculator will be able to devise a system which will cover the case when a function of an angle in any of the other three quadrants is required.

The RPN examples will suit the Hewlett-Packard (HP) family of calculators; they are based on the HP-25 keyboard, being in the middle of the price range, and operators of, say, the HP-67 will readily recognize the occasions when a different shift key should be used for a listed operation; for example, h π for g π. The HP-25 also provides a 49-step programming facility for repetitive calculations, while the HP-19C and HP-29C provide 98 steps and the HP-67 and HP-97 provide 224 steps. The Appendix contains fully documented and tested programmes (based on some of the chapter topics) written especially for these calculators. Users of the more advanced HP-67 and HP-97 models will be able to record these programmes on magnetic storage cards for immediate use. Users of Texas Instruments and other comparable models will be able to adapt these programmes to match their own keyboard facilities and logic.

Much time can be saved in this manner when a series of similar reductions has to be processed, as the calculator will be working at its maximum speed for most of the time, under the control of the programmed instructions, as a mini-computer in its own right. In addition, such programmes greatly reduce the chances of mistakes arising through operator fatigue or incorrect keying.

Although I have selected the programmable HP-25 and HP-67 for my own use, and the RPN instructions in this book must obviously reflect that choice, I should take this opportunity to stress the fact that this is an extremely competitive market with advances constantly being made, and models from other manufacturers must not be regarded as inferior simply because they have not been referred to specifically in the text.

Remember, above all, that it is answers we require. And we want them quickly and reliably. The calculator itself is only a tool which we can employ in order to meet those objectives. It should go without saying, therefore, that we must guard against getting sidetracked by questions of elegance of the method of solution, or becoming champions of the irrelevant claims and counterclaims of one product or logic system in preference to another.

I am indebted to M Jean Meeus of the *Vereniging voor Sterrenkunde* (Belgium) and of the British Astronomical Association for reading the manuscript of this book, and for his invaluable comments and helpful suggestions. With his kind permission, several of his programmes for the HP-67 have been included in the Appendix.

—A.J. Kent, 1978

Introduction

Each chapter of this book deals with a separate theme: time, position, and so on. Some chapters cover more than one topic under the general theme; in these cases the topics are numbered serially.

Worked examples are numbered to correspond with the topic number, followed by a letter, A or B. The A method for each example relates to the working to be employed with a simple type of scientific calculator, with single memory facility and 8-digit display, using algebraic logic. The B method shows the more sophisticated keying which can be employed with a calculator equipped with up to 7 user-addressable storage memories and 10-digit display, using Reverse Polish Notation (RPN). The same problem is solved as in the A method.

In most cases it will be found that the B method gives the more accurate result, or is shorter in execution. Accuracy limits, where relevant, are quoted in the topic heading. The layout of the A and B examples follows the format:

Numbered step-by-step statement of the solution to the problem	Keyboard entries*	Blank for user's modifications	Explanatory notes and any manual computation†

Throughout the book, unless noted to the contrary, symbols and Greek letters are used in exactly the same sense as in the *Astronomical Ephemeris* (*AE*) and as defined therein or in its *Explanatory Supplement.*

Some of the topics covered often occur in practice as cases where a large number of identical calculations have to be processed serially. For example, the reduction for precession from one epoch to another may have to be computed for a number of stars at the same time. The text indicates, where appropriate, that programmes

* Entries in this column enclosed in square brackets [ ] show figures which must be keyed in to solve the worked example.

† Entries in this column enclosed in round brackets ( ) are for check purposes only. They show the displayed contents of the X-register at significant points during the working of the example. Once the method of working has been proved with a particular type of calculator these entries can be ignored.

especially written for this purpose are included in the Appendix at the back of the book.

In a book of this nature it is clearly impossible to include programmes for every single calculation the reader may wish to perform. For this reason, blank pages have been left at the end of each chapter which the reader may use for his own extension of that chapter—or for any other notes which he may wish to add.

1 Time

Topic 1 Greenwich Mean Sidereal Time (with note about Greenwich Apparent Sidereal Time)

Topic 2 Local Sidereal Time

Topic 3 Local Mean Time

1 **To calculate Greenwich Mean Sidereal Time (GMST) at 0^h Universal Time (UT) on January 0 of any year after 1900, correct to $\pm 0^s.1$ with a simple 8-digit algebraic calculator (Method 1A); correct to $\pm 0^s.001$ (for any day of the year) with a 10-digit RPN calculator (Method 1B).**

Introduction: This value is readily available from the *Astronomical Ephemeris* (*AE*) for the current year, and appears as a constant (for that year) in many calculations involving time. In normal circumstances it will not therefore need to be calculated. However, on occasion it will be desirable to compute its value in advance of publication of the *AE*, or perhaps for a past year for which the appropriate *AE* is no longer held.

The equation: $6^h\,38^m\,45^s.836 + 8\,640\,184^s.542\,T + 0^s.092\,9\,T^2$ (1.1)
where T is the number of Julian centuries of 36 525 days of UT elapsed since 12^h UT on 1900, January 0.

Note: For the Apparent Greenwich Sidereal Time a correction for the equation of the equinoxes ($\Delta\psi \cos\epsilon$) would have to be added to the result given by Eqn. 1.1. $\cos\epsilon$ is easily evaluated, but there are 69 terms in $\Delta\psi$, so it is not so readily determined. (But see Chapter 9 for a reliable approximation.) The daily value of the equation of the equinoxes for current and past years is obtained, if required, from Column 5 on pp 12–19 of the *AE*. For example, the value for 1978, January 0 is $+0^s.226$. Many observers choose to ignore this difference.

Further information: *Explanatory Supplement to the AE*, pp 43, 72, 75, 84, 85 and 92.

Example 1: What is the Mean Sidereal Time at Greenwich at 0^h UT on 1978, January 0?

Method 1A

	MC	Clear memory
1. Subtract 1900 from desired year	**[1978]**	
	−	
	1900	
2. Convert to days	×	
	365	
3. Add number of leap days during period	+	(1900 was not a leap year)
	[19]	
4. Deduct $0^d.5$	−	Count starts from 12^h 1900, January 0
	0.5	
5. Convert to Julian centuries	÷	
	3 652.5	We cheat in order to obtain an extra decimal place
6. Note T for Step 16	=	(0.779 972 62). We have moved the decimal point one place to the left to compensate for the cheat in Step 5
7. Store display in memory	**M+**	(7.799 726 2)
8. Compute second term	×	
	8 640 184	
	=	
	←→	
	XM	Transfer X display to memory and vice versa
	×	
	0.542	
	=	
	M+	Add to figure already in memory, and recall total
	MR	
Compensate for cheat, Step 5	÷	
	10	
	=	(6 739 107.3)
Note decimal part of display and carry forward to Step 15.		$0^s.3$
Deduct the decimal part	−	
	[0.3]	
	=	
9. Reduce integer to days	**MC**	
	M+	
	÷	
	86 400	
	=	($77^d.998\ 923$)
10. Multiply integral number of days by 86 400 and subtract from the number of seconds stored in the memory	[77]	
	×	
	86 400	
	=	
	←→	
	XM	
	−	
	MR	
11. Divide by 86 400 for the fraction of a day	÷	
	86 400	
	=	($0^d.998\ 923$)

Method 1A *continued*

12. Convert to hours	×	
	24	
	=	($23^{h}.974\ 166$)
13. Note hours; subtract integral number of hours and convert fraction to minutes	–	23^{h}
	[23]	
	×	
	60	
	=	($58^{m}.449\ 96$)
14. Note minutes; subtract integral minutes and convert fraction to seconds	–	58^{m}
	[58]	
	×	
	60	
	=	($26^{s}.997\ 6$)
15. Add fraction of seconds from Step 8; note seconds	+	
	[0.3]	
	=	$27^{s}.298$
16. Square T from Step 6 and compute third term	**[0.779 972 6]**	
	×	
	=	
	×	
	0.092 9	
Note third term	=	$0^{s}.057$

	h	m	s
17. Sum constant and evaluated second and third terms	6	38	45.836
	23	58	27.298
			0.057
	30	37	13.191
18. If necessary, deduct 24 hours	–24		
	6	37	13.2

Result 1A. GMST at 0^{h} on 1978, January 0 is $6^{h}\ 37^{m}\ 13^{s}.2$. The *AE* gives $6^{h}\ 37^{m}\ 13^{s}.280$. The computational error using the simple 8-digit algebraic-logic calculator is, in this case, $-0^{s}.08$, within the claimed limit of accuracy.

Method 1B

1. Enter t	**[1978]**	
	↑	'Enter' (replicate in Y register so that x can be overwritten by a new x)
	1900	
	–	
2. Convert to days	**365**	
	×	
3. Add number of leap days	**[19]**	(1900 was not a leap year)
	+	
4. Deduct $0^{d}.5$	**0.5**	Count starts from 12^{h} 1900, January 0
	–	
5. Convert to centuries	**36 525**	
	÷	(0.779 972 622) T in Julian centuries
6. Store for Step 18	**STO 0**	HP-25 addressable memories are numbered 0 to 7 inclusive

Method 1B *continued*

Step	Keystrokes	Comments
7. Compute second term	**8 640 184.542**	
	×	
	g FRAC	Discard integral x
	STO 2	For Step 16
	f last x	
	f INT	Truncate decimal fraction
	STO 1	
8. Reduce integral seconds to days	**86 400**	
	÷	
9. Multiply integral days by 86 400 and subtract from number of seconds in R1	**f last x**	i.e., 86 400
	$x \longleftrightarrow y$	
	f INT	
	×	
	RCL 1	
	$x \longleftrightarrow y$	
	−	
10. Divide by 86 400 for fraction of day	**86 400**	
	÷	(0^{d}.998 923 611)
11. Convert to hours	**24**	
	×	(23.974 166 67)
12. Note integral hours		23^{h}
23. Convert fraction to minutes	**g FRAC**	Doing this the long way
	60	as shown gives greater
	×	accuracy than if f H.MS
		(58.450 000 20)
14. Note minutes		58^{m}
15. Convert fraction to seconds	**g FRAC**	
	60	
	×	
16. Add fractional seconds from Step 7; note seconds	**RCL 2**	
	+	(27.387 012 00)
		27^{s}.387
		Value of second term is: 23^{h} 58^{m} 27^{s}.387
18. Square T and compute second term	**RCL 0**	
	↑	'Enter'
	×	
	0.092 9	
	×	(0.056 5)
19. Note third term		0^{s}.057
20. Sum constant with evaluated second and third terms		h m s
		6 38 45.836
		23 58 27.387
		0.057
		30 37 13.280
21. If necessary, deduct 24 hours		−24
		6 37 13.280

Result 1B. Greenwich Mean Sidereal Time at 0^{h} UT on 1978, January 0 is 6^{h} 37^{m} 13^{s}.280. The *AE* gives exactly the same value, so the computational error in this case is zero. By using a 10-digit multi-memory RPN-logic calculator, such as the HP-25, the claimed accuracy is easily achieved.

Additional notes on Method 1B: If the user is content with a slightly lower degree of accuracy, say to $\pm 0^s.01$, then it is possible to reduce the number of steps in the calculation, as follows:

1. Perform Steps 1–6				
2. Then:	**8 640 184.542**			
	↑			
	86 400			
	÷			
	×			
	g FRAC			
	24			
	×			
	f H.MS	(23.582 738)		
3. Note second term in H.MS format		h	m	s
		23	58	27.38
4. Do Steps 18–21				0.06
		6	38	45.84
		30	37	13.28
		–24		
		6	37	13.28

The result now is $6^h\ 37^m\ 13^s.28$. The *AE* value, rounded to two places of decimals, is $6^h\ 37^m\ 13^s.28$. The error is still nil.

Method 1B is not restricted to the calculation of GMST at 0^h UT on January 0. It can also be used to find GMST at 0^h UT on any day. The *Explanatory Supplement to the AE*, p. 84, gives a solution for 0^h UT on 1960, March 7: $10^h\ 58^m\ 50^s.971$. For practice, use Method 1B to solve the same example.

In this case, the number of leap days to add in Step 3 is 14. Between Steps 3 and 4, add $31 + 29 + 7 = 67$ days for the period January 0 to March 7 (1960 was a leap year). You should obtain $10^h\ 58^m\ 50^s.972$. The computational error is $+0^s.001$.

Finally, to construct an ephemeris to show GMST at 0^h UT for any year after 1900, see the programmes especially written for the HP-25 in the Appendix at the back of the book (Programmes 1 and 2). After keying in the programme, one is able to progress day-by-day through the year by simply pressing the Start key for each successive day's calculation, thus saving an immense time in the compilation and avoiding any danger of operator keying errors. A similar programme for the HP-67, offering various options, is also included (Programme 5). This is valid from the year 1582.

Programmes 3 and 4 will also be found useful for finding the sidereal time other than at 0^h UT.

2 To calculate Local Sidereal Time (LST) at any observer's position, at any time, to an accuracy of $\pm 0^s.1$. For a series of similar calculations, see the programmes in the Appendix (Programmes 3 to 5).

Introduction: LST is required for setting the hour circle of an equatorial mount. It is unique to the observer's meridian, and is therefore dependent upon the ST at

Greenwich plus an adjustment for the difference in longitude between the observer's position and 0°. Local Standard Time is the clock time shown in the observer's time zone. Time zones are 15° of longitude in width, centred on the meridians at 0°, 15°, 30°, 45°, etc., the Earth thus being divided into 24 zones of one-hour time differences.

The zone centred on 0°, the meridian at Greenwich, uses Greenwich Mean Time (GMT), now known astronomically as Universal Time (UT), from which the time in other zones is derived.

When calculating the LST we shall always proceed from the clock time in the observer's own time zone.

The equation: $LST = 0.002\,737\,909\,3\,(24d + x) + x + x_1 + \lambda$ (1.2)

where d = the number of days since January 0 of year

x = the equivalent UT of local-zone time (i.e., GMT + 0, CET − 1, EST + 5, CST + 6, MST + 7, PST + 8, etc.)

x_1 = GMST at 0^h UT on January 0 of current year, from Topic 1 or the *AE*, expressed in decimal hours

λ = longitude (degrees) ÷ 15 (+ if E of Greenwich, − if W of Greenwich)

Further information: The *AE*, annually, on or about p 536; D. Menzel, *A Field Guide to the Stars and Planets*, Chap. XIII, pp 322–326; Mayall & Mayall, *Skyshooting—Photography for Amateur Astronomers*, Chap. 13; D. McNally, *Positional Astronomy*, Chap. V.

Example 2: What is the LST at 9.30 pm EST (2130 hrs) on 1978, May 15, at Cambridge, Mass., where the observer's longitude is 71° 07′ 30″ W?

The standard time-zone difference is −5 hours, so to convert EST to UT, add 5 hours. In this example:

$d = 135$

$x = 21.5 + 5 = 26.5$ (i.e., 0230 UT May 16)

$x_1 = 6.620\,355\,556$ ($6^h\,37^m\,13^s.280$ for 1978), from the *AE* or Topic 1

$\lambda = -4.741\,666\,667$

Method 2A

1. Enter x, clock time + integral hours to convert to UT	**[26.5]**
2. Is local Daylight Saving Time or BST in operation? If so, deduct 1 hour	**[no operation]**
3. Store	**M+**
4. Enter d and convert to hours	**[135]** × **24**
5. Add x	+ **MR**
6. Multiply by hourly rate of gain of ST over MT	× **2.7379093** ÷ **1000**

Method 2A *continued*

7. Add x	+ MR	
8. Add x_1 on January 0 of year	+ [6.620 355 6]	
9. Add. λ in hours	+ [4.741 666 7] [CS]	λ is negative; the Step-9 entries will be constant for a fixed observatory
	=	(37.322 069)
10. If necessary, deduct ·24 hours	[– 24 =]	(13.322 069)
11. Note integral hours		13^h
12. Subtract hours and convert the fraction into minutes	– [13] × 60 =	
13. Note integral minutes		19^m
14. Deduct minutes and convert to seconds	– [19] × 60 =	
15. Note seconds		$19^s.45$ LST = $13^h\ 19^m\ 19^s.45$

Result 2A. The Local Sidereal Time at Cambridge, Mass., at 9.30 pm EST on 1978, May 15 is $13^h\ 19^m\ 19^s.45$. (To obtain the apparent LST, add the equation of the equinoxes.)

Method 2B

1. Fix display and enter x	f FIX 6 [26.5]	For 6 places of decimals
	↑	'Enter'
2. Is Daylight Saving Time (BST in the UK) in operation? If so, deduct 1 hour	[no operation]	
3. Store	STO 0	
4. Enter d and convert to hours	[135] ↑	'Enter'
	24 ×	
5. Add x	RCL 0 +	
6. Complete first term	0.027 379 093 × 10 ÷	(8.943 381)
7. Add x	RCL 0 +	

Method 2B *continued*

8. Add x_1	**[6.620 355 556]**	
	+	
9. Add λ	**[4.741 666 667 CHS]**	λ is negative
	+	(37.322 070)
10. If necessary, deduct 24 hours	**[24 −]**	
11. Convert to hours, minutes and seconds	**f H.MS**	(13.19 19 45) Read as: $13^h\ 19^m\ 19^s.45$

Result 2B. The Local Sidereal Time at Cambridge, Mass., at 9.30 pm EST on 1978, May 15 is $13^h\ 19^m\ 19^s.45$. (To obtain the apparent LST, add the equation of the equinoxes.)

For further practice, try the following:

(a) What was the LST at Cleveland, Ohio, 81° 45′ W, at $12^h\ 30^m\ 21^s$ EST on 1964, May 29? In this case, $d = 150$, $x = 17.505\ 833\ 33$, $x_1 = 6.580\ 318\ 889$, $\lambda = -5.450$. (Check first to see if you agree these values.)

(b) What was the Apparent Local Sidereal Time at $3^h\ 44^m\ 30^s$ am CST on 1976, July 7 for an observer located at 85° 15′ W, given $x_1 = 6.586\ 474\ 722$ and that the equation of the equinoxes was $+0^s.71$?

(c) What was the Apparent Local Sidereal Time at Rainham, Kent, 0° 35′ 54″.4 E, at 12.30 am BST on 1976, September 22, given $x_1 = 6.586\ 474\ 722$ and that the equation of the equinoxes was $+0^s.62$?

(d) Imagine, as a visitor, you were granted permission temporarily to erect a small telescope in the grounds of Siding Springs Observatory, New South Wales, Australia. You did not have the current ephemeris, but wanted to know the LST at 8 pm clock time on 1976, October 1. Fortunately, you knew the longitude of Siding Springs is 149° 04′.0 E; also, you had remembered to take this book and your pocket calculator with you. What was the LST?

Your answers should be:

(a) $4^h\ 32^m\ 26^s.0$;
(b) $23^h\ 05^m\ 27^s.0$;
(c) $23^h\ 36^m\ 14^s.35$;
(d) $20^h\ 37^m\ 18^s.59$.

If your solutions to (b) and (c) were incorrect, check that you remembered to add the equation of the equinoxes at the end of the calculation, and that you used, in (b), $d = 189$, $x = 9.741\ 666\ 667$, $\lambda = -5.683\ 333\ 333$; and, in (c), $d = 266$, $x = 0.5$, $\lambda = +0.039\ 896\ 296$; also in (c), that in Step 2 you keyed in −1 to correct for BST.

3 To calculate Local Mean Time, to an accuracy of $\pm 0^s.01$.

Introduction: Local Mean Time (LMT), like LST, is unique to the observer's position in longitude, and is the mean solar time at that place. Because of the confusion which would otherwise arise from a multiplicity of local times it is more convenient for civil time-keeping purposes to use the standard time zones. LMT for an observer located east of his standard time meridian will be in advance of clock time by 4^m in respect of each degree of longitude, 4^s for each minute of longitude, etc. An observer west of his standard time meridian will have LMT running later than zone time by the same amounts.

The equation: $\text{LMT} = x + (\lambda_1 - \lambda_2)$ (1.3)

where x = clock time (standard time for the observer's zone), expressed in decimal hours

λ_1 = longitude of standard time meridian of x, expressed in hours, positive if W of Greenwich, negative if E

λ_2 = longitude of observer, expressed in hours, positive if W of Greenwich, negative if E

$\lambda_1 - \lambda_2$ will be a constant for a fixed observing location. Once evaluated for a particular observatory, it need not be calculated again.

Further Information: D. Menzel, *A Field Guide to the Stars and Planets*, Chap. XIII, pp 322–324; G. D. Roth, *Astronomy—A Handbook*, 6.3.1.1., pp 171–172.

Example 3: An observer is located at 77° 10′ W. What is his LMT at 8.00 pm EST?

In this example, x = 20 (hours)

λ_1 = 5 (hours) (75° ÷ 15)

λ_2 = 5.144 444 (hours) (77°.166 667 ÷ 15)

Method 3A

1. Enter λ_1	**[5]**	λ_1 and λ_2 both positive
2. Subtract λ_2	−	
	[5.144 444]	
3. Add clock time, x	+	
	[20]	
	=	
4. If necessary, deduct 24 hours	**[no operation]**	
5. Note integral hours		19^h
6. Subtract hours and convert remainder to minutes	−	
	[19]	
	×	
	60	
7. Note minutes	=	51^m
8. Subtract minutes and convert remainder to seconds	−	
	[51]	
	×	
	60	
9. Note seconds	=	$20^s.0$
		LMT = $19^h\ 51^m\ 20^s.0$

Result 3A. LMT for an observer at 77° 10′ W at 8.00 pm EST is $19^h\ 51^m\ 20^s$. Note: if this location is a permanent one, observers would key in $\lambda_1 - \lambda_2$, previously evaluated as −0.144 444, and start at Step 3.

Method 3B

In this method there is no need to evaluate λ_1 and λ_2 first, unless this is a permanent location for the observer, in which case he would input (for this example) 0.144 444, CHS, ↑ , and start at Step 5.

1. Fix 6 decimal places	**f FIX 6**	
2. Enter λ_1	[**75**]	λ_1 and λ_2 both positive
	↑	'Enter'
3. Longitude of observer in D.MS format; convert to decimal degrees and subtract	[**77.10**] **g → H** **–**	
4. Convert to hours	**15** ÷	
5. Enter clock time (zone time) in H.MS format, convert to decimal hours and add	[**20**] **g → H** +	
6. Convert to hours, minutes and seconds	**f H.MS**	(19.51 20 00) Read as: $19^h\ 51^m\ 20^s.00$

Result 3B. LMT for an observer at 77° 10′ W at 8.00 pm EST is $19^h\ 51^m\ 20^s.0$.

For further practice, try the following:

(a) What is the LMT for an observer situated at 0° 35′ 54″.4 E, at 8.10 pm (2010 UT)? (Note: λ_2 will be negative.)

(b) An observer is located at 20° 30′ E. What is his LMT at 2215 CET? (Both λ_1 and λ_2 are negative.)

(c) What are the constants, expressed in minutes and seconds, to be applied to clock time at observatories located at (*i*) 92° 15′ 21″.2 W, (*ii*) 11° 20′ 15″.8 E, in order to determine LMT? (That is, $\lambda_1 - \lambda_2$ expressed in minutes and seconds of time.)

(d) An observer calculates that the transit time of a star is due at Greenwich at $22^h\ 15^m\ 10^s.2$ UT on a certain day. At his observatory he times the actual transit at $22^h\ 41^m\ 22^s.2$ UT. What is his longitude, to the nearest minute?

Your answers should be:

(a) $20^h\ 12^m\ 23^s.63$;
(b) $22^h\ 37^m\ 00^s$;
(c) (*i*) $-9^m\ 01^s.41$, (*ii*) $-14^m\ 38^s.94$;
(d) 6° 33′ W.

NOTES

2 Precessional Constants for Selected Epochs

Topic 1 Angles defining total precession from the equator and equinox of epoch t_0 to the equator and equinox of a later epoch t: ζ_0, z, θ, $\sin\theta$, $\tan\frac{1}{2}\theta$.

Topic 2 Mean obliquity of the ecliptic, ϵ.

Topic 3 Annual general precession, p; annual precession in RA, m; annual precession in dec., n.

1 **To calculate ζ_0, z, θ, $\sin\theta$ and $\tan\frac{1}{2}\theta$ for any year in relation to any catalogue epoch, or vice versa, with an accuracy of ± 1 in the last digit with a simple 8-digit algebraic calculator (Method 1A); correct to 9 decimal places with a 10-digit RPN calculator (Method 1B).**

Introductions These values, which are required for the reduction of star positions from one epoch to another, are obtainable from the *AE* for the current year; they give the constants which apply to reductions from that particular epoch to the equator and equinox of 1950.0. However, it will often be required to reduce star positions to and from other epochs, and also to evaluate the constants in advance of publication of the *AE* for a particular year. Users of the HP-67 programme in the Appendix (Programme 11) for reductions for precession will find that ζ_0, z and θ are calculated automatically. If only the constants are required, see Programme 6.

The equations:

$$\zeta_0 = (23\,042''.53 + 139''.73\,\tau + 0''.06\,\tau^2)\,T + (30''.23 - 0''.27\,\tau)\,T^2 + 18''.00\,T^3 \qquad (2.1)$$

$$z = \zeta_0 + (79''.27 + 0''.66\,\tau)\,T^2 + 0''.32\,T^3 \qquad (2.2)$$

$$\theta = (20\,046''.85 - 85''.33\,\tau - 0''.37\,\tau^2)T + (-42''.67 - 0''.37\,\tau)T^2 - 41''.80\,T^3 \qquad (2.3)$$

where $\tau = \dfrac{t_0 - 1900.0}{1000}$

$T = \dfrac{t - t_0}{1000}$

t_0 = epoch of earlier equinox ⎫ beginning of Besselian
and t = epoch of later equinox ⎭ solar year.

Equations 2.1 to 2.3 give the values for ζ_0, z and θ to be used in reductions from an earlier epoch t_0 to a current or later epoch t, where the equatorial coordinates for epoch t_0 are known and it is desired to find the positions for the equator and equinox of epoch t. In other words, the reduction goes $t_0 \rightarrow t$. If the reverse is desired—that is, the position for the epoch t is known and the reduction goes $t \rightarrow t_0$—then, after solving the equations in the normal way, for ζ_0 use $-z$, for z use $-\zeta_0$, and change the sign for θ. An example of such a reduction is given in Chapter 3, Topic 3.

Methods 1A and 1B give the values in decimal degrees.

Further information: Any current *AE*, on or near p 534, under the heading Precessional Constants; also p 9; *Explanatory Supplement to the AE*, p 30; Introduction to the *SAO Star Catalogue*, p *xii*; D. McNally, *Positional Astronomy*, 7.2, pp 162–164; H. K. Eichhorn, *Astronomy of Star Positions*; S. Newcomb, *A Compendium of Spherical Astronomy*, Chap. IX; Woolard and Clemence, *Spherical Astronomy*, Chap. 11.

Example 1: What are the values of ζ_0, z, θ, $\sin\theta$ and $\tan\frac{1}{2}\theta$ to be employed in the reduction of star positions from the epoch 1950.0 to the equator and equinox of 1978.0?

Method 1A

1. Evaluate τ and τ^2	**MC**	Clear memory
	[1950]	t_0
	−	
	1900	
	÷	
	1000	
Note τ	=	$(0.05) = \tau$
	×	
	=	
Note τ^2	**M+**	$(0.002\ 5) = \tau^2$
2. Evaluate T, T^2 and T^3	**[1978]**	t
	−	
	[1950]	t_0
	÷	
	1000	
Note T	=	$(0.028) = T$
	×	
Note T^2	=	$(0.000\ 784) = T^2$
Note T^3	=	$(0.000\ 021\ 9) = T^3$
3. First term of Eqn. 2.1	**MR**	τ^2
	×	
	0.06	
	=	
	MC	
	M+	
Enter τ	**[0.05]**	
	×	
	139.73	
	+	
	23 042.53	

Method 1A *continued*

	+	
	MR	
	×	
Enter T	**[0.028]**	
	=	
	MC	
	M+	(645.386 44) First term
4. Second term of Eqn. 2.1	**[0.05]**	
	×	
	0.27	
	CS	Change sign
	+	
	30.23	
	×	
Enter T^2	**[0.000 784]**	
	=	(0.023 689 7) Second term
Third term	**M+**	
	18	
	×	
Enter T^3	**[0.000 021 9]**	
	=	(0.000 394 2) Third term
	M+	
5. Convert ζ_0 to degrees	**MR**	
	÷	
	360	
	=	(1.792 806 9)
Move decimal point one place to left and note ζ_0 to 7 places		$\zeta_0 = 0^\circ.179\ 280\ 7$
6. Evaluate z, Eqn. 2.2	**[0.05]**	τ
	×	
	0.66	
	+	
	79.27	
	×	
Enter T^2	**[0.000 784]**	
	=	
	M+	Add to ζ_0 in store
Enter T^3	**[0.000 021 9]**	
	×	
	0.32	
	=	
	M+	
	MR	
	÷	
	360	
	=	(1.792 979 6)
Move decimal point one place to left and note z to 7 places		$z = 0^\circ.179\ 298\ 0$
7. θ, first term of Eqn 2.3	**0.37**	
	CS	
	×	
Enter τ^2	**[0.0025]**	

Method 1A *continued*

	=	
	MC	
	M+	
	85.33	
	CS	
	×	
Enter τ	**[0.05]**	
	+	
	20 046.85	
	+	
	MR	
	×	
Enter T	**[0.028]**	
	=	(561.192 35) First term
	MC	
	M+	
8. Second term of Eqn. 2.3	**0.37**	
	CS	
	×	
Enter T	**[0.028]**	
	+	
	42.67	
	CS	
	×	
Enter T^2	**[0.000 784]**	
	=	(–0.033 461 4) Second term
	M+	
9. Third term of Eqn. 2.3	**41.80**	
	CS	
	×	
Enter T^3	**[0.000 021 9]**	
	=	(–0.000 915 4) Third term
	M+	
10. Convert θ to degrees	**MR**	
	÷	
	360	
	=	(1.558 772 1)
Move decimal point one place to left and note θ to 7 places		$\theta = 0^\circ.155\,877\,2$
11. Evaluate $\sin\theta$ and $\tan\frac{1}{2}\theta$	÷	
	10	
	=	
	MC	
	M+	
Note $\sin\theta$	**f sin**	$\sin\theta = 0.002\,720$
	MR	
	÷	
	2	
	=	
Note $\tan\frac{1}{2}\theta$	**f tan**	$\tan\frac{1}{2}\theta = 0.001\,360$

Result 1A. The values of the precessional constants required to reduce star positions from epoch 1950.0 to the mean equator and equinox of 1978.0 are: $\zeta_0 = 0°.179\,280\,7$, $z = 0°.179\,298\,0$, $\theta = 0°.155\,877\,2$, $\sin\theta = 0.002\,720$, $\tan\frac{1}{2}\theta = 0.001\,360$. The values listed in the *AE* for 1978 are: $\zeta_0 = 10'\,45''.41$ ($= 0°.179\,280\,709$)*, $z = 10'\,45''.47$ ($= 0°.179\,297\,981$)*, $\theta = 9'\,21''.16$ ($= 0°.155\,877\,202$)*, $\sin\theta = 0.002\,720\,56$, $\tan\frac{1}{2}\theta = 0.001\,360\,28$.

The results from the simple calculator are correct to seven decimal places, within the claimed limits. The values of $\sin\theta$ and $\tan\frac{1}{2}\theta$ are correct to the six places of decimals given by the calculator used.

Method 1B

	f FIX 9	Display 9 decimal places
1. Enter latest epoch, t	**[1978]**	
	↑	'Enter'
2. Enter earlier epoch, t_0	**[1950]**	
	STO 4	
3. Find constants	−	
	1000	
	÷	
	STO 2	T
	RCL 4	
	1900	
	−	
	1000	
	÷	
	STO 3	τ
	RCL 2	
	3	
	f y^x	
	18	
	×	
	STO 0	
	RCL 2	
	g x^2	
	RCL 3	
	0.27	
	×	
	CHS	
	30.23	
	+	
	×	
	STO + 0	
	RCL 3	
	g x^2	
	0.06	
	×	
	RCL 3	
	139.73	
	×	
	+	
	23 042.53	

* Decimal equivalents from Method 1B.

Method 1B *continued*

+
RCL 2
×
STO + 0
RCL 0 (645.410 551 1) ζ_0 in seconds
3 600
÷
STO 5
RCL 2
3
f y^x
0.32
×
STO 0
RCL 2
g x^2
RCL 3
0.66
×
79.27
+
×
STO + 0
RCL 0
3 600
÷
RCL 5
+
STO 6 (0.179 297 981) z in degrees
RCL 2
3
f y^x
41.80
×
CHS
STO 0
RCL 2
g x^2
RCL 3
0.37
×
CHS
42.67
CHS
+
×
STO + 0
RCL 3
g x^2
0.37
×
CHS
RCL 3
85.33

Method 1B *continued*

	×	
	CHS	
	+	
	20 046.85	
	+	
	RCL 2	
	×	
	STO + 0	
	RCL 0	(561.157 926 8) θ in seconds
	3 600	
	÷	
	STO 7	
4(a). If reduction is from t_0 to t (e.g. 1950 → 1978),		
Note ζ_0	**RCL 5**	$\zeta_0 = 0^\circ.179\ 280\ 709$
Note z	**RCL 6**	$z = 0^\circ.179\ 297\ 981$
Note θ	**RCL 7**	$\theta = 0^\circ.155\ 877\ 202$
	2	
	÷	
Note $\tan\frac{1}{2}\theta$	**f tan**	$\tan\frac{1}{2}\theta = 0.001\ 360\ 286$
	RCL 7	
Note $\sin\theta$	**f sin**	$\sin\theta = 0.002\ 720\ 567$
4(b). If reduction is from t to t_0 (e.g. 1978 → 1950),		
	RCL 6	
Note ζ_0	**CHS**	$\zeta_0 = -0^\circ.179\ 297\ 981$
	RCL 5	
Note z	**CHS**	$z = -0^\circ.179\ 280\ 709$
	RCL 7	
Note θ	**CHS**	$\theta = -0^\circ.155\ 877\ 202$
	2	
	÷	
Note $\tan\frac{1}{2}\theta$	**f tan**	$\tan\frac{1}{2}\theta = -0.001\ 360\ 286$
	RCL 7	
	CHS	
Note $\sin\theta$	**f sin**	$\sin\theta = -0.002\ 720\ 567$

Result 1B. The values of the precessional constants required to reduce star positions from epoch 1950.0 to the mean equator and equinox of 1978.0 are: $\zeta_0 = 0^\circ.179\ 280\ 709$, $z = 0^\circ.179\ 297\ 981$, $\theta = 0^\circ.155\ 877\ 202$, $\sin\theta = 0.002\ 720\ 567$, $\tan\frac{1}{2}\theta = 0.001\ 360\ 286$. Compare with the values quoted in the *AE* for 1978: $\zeta_0 = 10'\ 45''.41$ (the value for ζ_0 in seconds noted in the remarks column above was $645.410\ 551\ 1 = 10'\ 45''.41$), $z = 10'\ 45''.47$ (not obtainable from our remarks column), $\theta = 9'\ 21''.16$ (noted in our remarks column, in seconds, $561.157\ 926\ 8 = 9'\ 21''.158$), $\sin\theta = 0.002\ 720\ 56$, $\tan\frac{1}{2}\theta = 0.001\ 360\ 28$.

We can therefore modestly claim a slightly greater accuracy for our values over those quoted in the *AE*. Method 1B demonstrates the power of the more advanced type of calculator, exploiting memory storage facilities to the full. It is only necessary to key in the two epochs in Steps 1 and 2, and the calculation proceeds without pause until the final stage at Step 4. No intermediate values have to be written down, and no manual computation has to be done. We also have an alternative ending to suit the case when the reduction for precession is to be made in the opposite direction, from $t \rightarrow t_0$.

Methods 1A and 1B are suitable for *any* two epochs: for example, from Auwers' catalogue of Bradley stars for 1755.0 → Argelander's *Bonner Durchmusterung* (*BD*) 1855.0, or *BD* 1855.0 → Boss' *General Catalogue* (*GC*) 1950.0, always provided that reliable values for the proper motions can be calculated (this is covered in Chapter 5).

For further practice, try the following:

(a) Evaluate ζ_0, z and θ in order to reduce observed positions of stars in 1980 (first corrected from apparent place to mean place for 1980.0) for comparison with the *SAO Star Catalogue*, epoch 1950.0. Give values in degrees.

(b) Reference to a reliably accurate position of a particular star at 1855.0 is made in contemporary observatory records. It is desired to compare this with the modern catalogue position listed in the *GC* (epoch 1950.0) in order to check the proper motions listed for 1950.0. What values of ζ_0, z and θ must be employed for the reduction? Give answers in degrees.

(c) What values of ζ_0, z and θ should be employed to convert positions given in a catalogue for 1950.0 to new catalogue positions for the epoch 2000.0? Give answers in degrees.

Your answers should be:

		ζ_0	z	θ
(a)	Method 1A	$\zeta_0 = -0°.192\,106\,8$	$z = -0°.192\,087\,0$	$\theta = -0°.167\,010\,6$
	1B	$\zeta_0 = -0°.192\,106\,823$	$z = -0°.192\,086\,995$	$\theta = -0°.167\,010\,536$
(b)	Method 1A	$\zeta_0 = 0°.607\,980\,9$	$z = 0°.608\,179\,6$	$\theta = 0°.528\,998\,4$
	1B	$\zeta_0 = 0°.607\,980\,940$	$z = 0°.608\,179\,667$	$\theta = 0°.528\,998\,522$
(c)	Method 1A	$\zeta_0 = 0°.320\,153\,8$	$z = 0°.320\,208\,8$	$\theta = 0°.278\,338\,1$
	1B	$\zeta_0 = 0°.320\,153\,784$	$z = 0°.320\,208\,867$	$\theta = 0°.278\,338\,106$

2 To calculate the mean obliquity of the ecliptic, ϵ, correct to $\pm 0''.01$ with an 8-digit algebraic calculator (Method 2A); correct to $\pm 0''.001$ with a 10-digit RPN calculator (Method 2B).

Introduction: Sinϵ and cosϵ feature in several types of astronomical calculation, including orbital position. As with the precessional constants covered in Topic 1 of this Chapter, the value for ϵ at the current epoch is always available from the *AE* for that year. But it will often be desirable to know its value for other years, and before publication of the relevant *AE*, together with its functions.

The equation: $\epsilon = 84\,428''.26 - 46''.845\,T - 0''.005\,9\,T^2 + 0''.001\,81\,T^3$ (2.7)
where T is measured in Julian centuries of 36 525 ephemeris days elapsed since 12^h ET on 1900, January 0.

Further information: *Explanatory Supplement to the AE*, pp 98–99, p 170, p 180; D. McNally, *Positional Astronomy*, pp 152–155.

Example 2: What was the mean obliquity of the ecliptic, in degrees, minutes and seconds, for epoch 1978.0?

Note: If, instead, mean obliquity of date is required for any particular day of the current year this can be obtained by adding, between Steps 4 and 5 of Methods 2A and 2B, the number of days elapsed since January 0.

Method 2A

1.	**MC**	Clear memory
Enter t	**[1978]**	
	−	
	1900	
2. Convert to days	×	
	365	
3. Add leap days since 1900	+	
	[19]	
4. Deduct $0^d.5$	−	Count starts from 12^h 1900, January 0
	0.5	
5. Convert to Julian centuries	÷	
	36 525	
	=	
6. Note T for Step 9	**M+**	0.779 972 6 (T in Julian centuries)
7. Second term	×	
	MR	
	=	
	←→	
	XM	T^2 in memory
	×	
	46.845	
	CS	
	=	
	←→	
8. Third term	**XM**	2nd term in memory
	×	
	0.0059	
	CS	
	=	
	M+	
9. Fourth term, enter T	**[0.779 972 6]**	T from Step 6
	y^x	
	3	
	=	(0.474 503) T^3
	×	
	0.001 81	
	=	
	M+	
10. Sum terms	**84 428.26**	
	+	
	MR	
	=	(84 391.720) ϵ in arc secs
11. Convert to degrees, minutes and seconds.	**MC**	
	M+	
	÷	
	3 600	
Note integral degrees	=	23°

Method 2A *continued*

12. Enter degrees	−	
	[23]	
	×	
	60	
Note minutes	=	26′
13. Enter minutes	−	
	[26]	
	×	
	60	
Note seconds	=	31″.72
14. If result is required in decimal degrees, for sinϵ cosϵ or tanϵ	←→	
	XM	
	MC	
	÷	
	3 600	
Note decimal degrees	=	23°.442 144
15. Evaluate sin, cos and tan	**M+**	
	f sin	sinϵ = 0.397 823
	MR	
	f cos	cosϵ = 0.917 462
	MR	
	f tan	tanϵ = 0.433 612

Result 2A. The mean obliquity of the ecliptic at epoch 1978.0 was 23° 26′ 31″.72 (23°.442 144). The values given in the 1978 *AE* are 23° 26′ 31″.719, 23°.442 144. The functions of ϵ compare as follows:

	Calculated	*AE*
sinϵ	0.397 823	0.397 822 84
cosϵ	0.917 462	0.917 462 25
tanϵ	0.433 612	0.433 612 22

The results are within the claimed accuracy.

Method 2B

1.	**f FIX 7**	Fix 7 places of decimals
Enter *t*	**[1978]**	
	↑	'Enter'
	1900	
	−	
2. Convert to days	**365**	
	×	
3. Add leap days since 1900	**[19]**	
	+	
4. Deduct 0ᵈ.5	**0.5**	
	−	
5. Convert to Julian centuries	**36 525**	
	÷	(0.779 972 6)
	STO 0	*T* in Julian centuries
6. Second term	**46.845**	
	CHS	
	×	
	STO 1	

Method 2B *continued*

7. Third term	**RCL 0**	
	g x^2	
	0.005 9	
	CHS	
	×	
	STO + 1	
8. Fourth term	**RCL 0**	
	3	
	f y^x	
	0.001 81	
	×	
	STO + 1	
9. Sum terms	**84 428.26**	
	STO + 1	
	RCL 1	(84 391″.719 45)
10. Convert to degrees, minutes and seconds	**f INT**	Truncate decimal fraction
	STO 2	
	3 600	
11. Note integral degrees	÷	23°
	f INT	
	3 600	
	×	
	RCL 2	
	$x \leftrightarrow y$	
	−	
	RCL 1	
	g FRAC	
	+	
	60	
Note minutes	÷	26′
	g FRAC	
	60	
Note seconds	×	31″.719 45
		ϵ = 23° 26′ 31″.719
12. If decimal degrees required, for functions of ϵ, then:	**RCL 1**	
	3 600	
	÷	23°.442 144 3
	STO 3	
	f FIX 9	
	f sin	sinϵ = 0.397 822 844
	RCL 3	
	f cos	cosϵ = 0.917 462 253
	RCL 3	
	f tan	tanϵ = 0.433 612 220

Result 2B. The mean obliquity of the ecliptic at epoch 1978.0 was:

	Calculated	*AE*
ϵ	23° 26′ 31″.719	23° 26′ 31″.719
	or 23°.442 144 3	23°.442 144
sinϵ	0.397 822 844	0.397 822 84
cosϵ	0.917 462 253	0.917 462 25
tanϵ	0.433 612 220	0.433 612 22

The values agree. Those calculated have the advantage of the extra decimal place. The results are thus well within the claimed accuracy limits.

3 To calculate p, m and n, being the annual general precession, annual precession in RA and annual precession in dec. respectively.

Introduction: These constants are employed in calculating precessional changes in RA and dec. They are most useful for short-term variations, where $t - t_0 < 10$ years, and the star is not near the celestial pole. See Chapter 3, Topic 2.

Where great accuracy is not required—e.g., for finding purposes—m and n can be employed to advantage over longer periods, providing the value is taken for the midpoint of the interval: $t_0 + \frac{t - t_0}{2}$. This method is used later for updating the approximate 1920 coordinates given in Webb's *Celestial Objects for Common Telescopes* to a current epoch, with accuracy sufficient to place the required object within the field of a finder telescope. See Chapter 3, Topic 1.

The equations:

$$p = 50''.256\,4 + 0''.022\,2\,T \tag{2.8}$$

$$m = 3^s.072\,34 + 0^s.001\,86\,T \tag{2.9}$$

$$n = 20''.046\,8 - 0''.008\,5\,T \tag{2.10}$$

where T is measured in centuries from 1900, with no distinction between the Julian and tropical century necessary. n is usually evaluated in both seconds of arc and in seconds of time.

Further information: Any current *AE*, on or near p 534 under the heading 'Precessional Constants'; *Explanatory Supplement to the AE*, pp 35–41, 169–170; S. Newcomb, *A Compendium of Spherical Astronomy*, Chap. IX; D. McNally, *Positional Astronomy*, Chap. VII; Woolard and Clemence, *Spherical Astronomy*, Chap 11.

Note: These calculations are quite straightforward and there is no need to run the risk of insulting the reader's intelligence by giving step-by-step workings as in previous topics.

Example: Evaluate m in seconds of time, n in both seconds of time and seconds of arc, for use in the reduction of the coordinates of a star from epoch 1950.0 to 1958.0.

The midpoint of the interval is 1954, so in Eqns. 2.9 and 2.10, $T = 0.54$.

We find: $m = 3^s.072\,34 + 0^s.001\,0 = 3^s.073\,34$

$$n = 20''.046\,8 - 0''.004\,6 = 20''.042\,2$$

$$\frac{n}{15} = 1^s.336\,15$$

For further practice, try:

(a) Evaluate p, m and n for 1977.0.
(b) Repeat for 1978.0.

Your results should be:

(a) $p = 50''.273\,5$, $m = 3^s.073\,77$, $n = 20''.040\,3$

(b) $p = 50''.273\,7$, $m = 3^s.073\,79$, $n = 20''.040\,2$

These results agree with the values given in the *AE*.

NOTES

3 Reduction for Precession

Topic 1 Approximate reduction of the 1920 coordinates quoted in the Dover paperback edition of Webb's *Celestial Objects for Common Telescopes*, for finding purposes.

Topic 2 Approximate reduction for precession and proper motion from the equator and equinox of epoch t_0 to the equator and equinox of epoch t.

Topic 3 Rigorous reduction for the effects of precession and proper motion from the equator and equinox of epoch t_0 to the equator and equinox of epoch t (or vice versa) when greater accuracy is required, and in particular for stars near the celestial poles.

Topic 4 Note on rotational geometry and an alternative method of rigorous reduction for precession, using the rectangular equatorial co-ordinates x, y, z.

1 **To update the 1920 coordinates quoted in the Dover paperback edition of Webb's** *Celestial Objects for Common Telescopes*, **with accuracy limited to that necessary to place the desired object near the centre of the field of a finder telescope.**

Introduction: This publication is still widely used by amateur astronomers. The latest coordinates given in the text are for 1920, in hours and minutes of RA, degrees and minutes of dec. In an appendix the approximate coordinates for 2000.0 are listed. Where faint objects are concerned, and setting circles are used on a properly-adjusted equatorial mount, it is convenient to update the 1920 co-ordinates to a current year.

Method 1A or 1B can be employed when new coordinates for a single star are required. Usually, however, the coordinates for a number of stars will be wanted for an observing session. In that event, users of Method 1B will prefer to save time by programming the computation if the calculator in use has such a facility. To this end a fully-documented programme for the HP-25 is included (page 160) in Appendix II. It is recommended that this programme be used when a number of approximate reductions from 1920 to a current epoch are required (Programme 7).

With a suitable readjustment of the programme constants, this programme may also be used for updating coordinates from another epoch, say 1950.0.

The equations: $\alpha = \alpha_0 + 0.004\,2\, t\,[3.07 + (1.3 \sin\alpha_0 \tan\delta_0)]$ (3.1)

$\delta = \delta_0 + 0.000\,28\, t\,(20.04 \cos\alpha_0)$ (3.2)

where α and δ = RA and dec., expressed in decimal degrees, at new epoch

α_0 and δ_0 = RA and dec., expressed in decimal degrees, at 1920

t = period in years from 1920 to new epoch

Within the brackets readers will recognize simplified values of m and n derived from Chapter 2, Topic 3.

Example 1: The 1920 coordinates for Delta Andromedae are given in Webb's *Celestial Objects* as $0^h\,35^m.0$, $+30°\,25'$. Ignoring the effect of proper motion, what are the approximate coordinates for 1977, sufficient for finding purposes? In this case $t = 57$.

Method 1A

1. Enter minutes of α_0 and convert to hours	[35.0]	
	÷	
	60	
2. Add hours of α_0 and convert to degrees	+	
	[0]	
	×	
	15	
	=	
3. Store; evaluate cos and sin; note for Steps 6 and 7	**M+**	
	f cos	(0.988 36) $\cos\alpha_0$
	MR	
	f sin	(0.152 12) $\sin\alpha_0$
4. Enter minutes of δ_0 and convert to degrees	[25]	
	÷	
	60	
5. Add degrees of δ_0	+	
	[30]	CS if Southern dec.
Note δ_0 for Step 7 and evaluate tan	=	(30°.416 66)
	f tan	
6. Solve for α	×	
Enter $\sin\alpha_0$	[0.152 12]	
	×	
	1.3	
	+	
	3.07	
	×	
Enter t	[57]	
	×	
	0.004 2	
	+	
	MR	
	÷	
	15	
Note integral hours	=	0^h
Deduct hours and convert to minutes	−	
	[0]	
	×	
	60	
Note minutes	=	$38^m.1$ $\alpha = 0^h\,38^m.1$

Method 1A *continued*

7. Solve for δ	**MC**	
Enter δ_0 from Step 5	**[30.416 666]**	CS if Southern dec.
	M+	
Enter $\cos\alpha_0$ from Step 5	**[0.988 36]**	
	$\times$	
	20.04	
	$\times$	
Enter t	**[57]**	
	$\times$	
	0.000 28	
	$+$	
	MR	
Note integral degrees	$=$	30°
Deduct degrees and	$-$	
convert to minutes	[30]	
	$\times$	
	60	
Note minutes	$=$	43′.967
		$\delta = +30° \, 44'$

Result 1A. The approximate coordinates for Delta Andromedae for 1977 are: $\alpha = 0^h\,38^m.1$, $\delta = +30° \, 44'$. The *AE* for 1977 gives $\alpha = 0^h\,38^m\,05^s.6$, $\delta = +30° \, 44' \, 07''$. The accuracy is quite good, and more than adequate for finding purposes; it is, in fact, good enough to place the star in the field of the main telescope. Take care, though, as the results for stars nearer the pole will not reach the same degree of accuracy, although they will still be good enough for the finder telescope.

Method 1B

	f FIX 4	Fix 4 places of decimals
1. Enter α_0 in H.MS format	**[0.35 00]**	The zeros are significant only inasmuch as they represent the number of seconds; e.g., $1^h\,36^m.3$ is $1^h\,36^m\,18^s$, each $0^m.1$ being 6 seconds, and would be entered here in H.MS format as 1.36 18
	g → H	
	15	
	$\times$	
	STO 0	
2. Enter δ_0 in D.MS format	[30.25]	CHS (change sign) if Southern dec.
	g → H	
	STO 1	
3. Enter t	[57]	
	STO 2	
4. Evaluate α	**RCL 1**	
	f tan	
	RCL 0	
	f sin	
	1.3	
	$\times$	
	3.07	
	$+$	
	RCL 2	

Method 1B *continued*

	×	
	0.004 2	
	×	
	RCL 0	
	+	
	15	
	÷	
Note α	**f H.MS**	0.38 08. Read as $0^{h}\ 38^{m}\ 08^{s}$ $\alpha = 0^{h}\ 38^{m}.1$
5. Evaluate δ	**RCL 0**	
	f cos	
	20.04	
	×	
	RCL 2	
	×	
	0.000 28	
	×	
	RCL 1	
	+	
Note δ	**f H.MS**	30.43 58. Read as $30^{\circ}\ 43'\ 58''$ $\delta = 30^{\circ}\ 44'$

Result 1B. The approximate coordinates for Delta Andromedae for 1977 are: $\alpha = 0^{h}\ 38^{m}.1$, $\delta = +30^{\circ}\ 44'$. The *AE* for 1977 gives $\alpha = 0^{h}\ 38^{m}\ 05^{s}.6$, $\delta = +30^{\circ}\ 44'\ 07''$. The accuracy is good, but see the word of caution with the result of Method 1A. Use the programme in Appendix II when several similar reductions have to be processed (Programme 7).

For further practice, try the following:

(a) *Celestial Objects* gives, for Eta Centauri, $14^{h}\ 30^{m}.4$, $-41^{\circ}\ 48'$ (1920). What are the approximate coordinates for 1977?

(b) M66, in Leo, has 1920 coordinates of $11^{h}\ 16^{m}.1$, $+13^{\circ}\ 26'$. Reduce these coordinates to epoch 1950.0.

(c) What will be the approximate coordinates of Zeta Herculis, given for 1920 in *Celestial Objects* as $16^{h}\ 38^{m}.3$, $+31^{\circ}\ 42'$, at epoch 2000.0?

Your answers should be:

(a) $14^{h}\ 32^{m}.6$, $-42^{\circ}\ 03'$. The *AE* for 1977 gives $14^{h}\ 34^{m}.0$, $-42^{\circ}\ 03'$.
(b) $11^{h}\ 17^{m}.8$, $+13^{\circ}\ 16'$. The *Atlas Cæli Catalogue* gives $11^{h}\ 17^{m}.6$, $+13^{\circ}\ 17'$.
(c) $16^{h}\ 40^{m}.8$, $+31^{\circ}\ 33'$. The *Celestial Objects* Appendix gives $16^{h}\ 41^{m}.3$, $+31^{\circ}\ 35'$.

The longer the intervening period the greater is the likely discrepancy. This is partly due to the effect of proper motion, which we have so far ignored, but is mainly a reminder that simplified equations are intended for use only over a relatively short period of time.

2 To calculate the effect of precession and proper motion on the mean equatorial coordinates α_0, δ_0, of a star between epoch t_0 (when the position and proper motion are known accurately) and a later epoch t, when $t - t_0 < 10$ years, and the star is not near either of the celestial poles, to an accuracy of $\pm 0^s.01$ in RA and $\pm 0''.01$ in dec.

Introduction: This is a fundamental problem which taxed the mathematical resources of the earliest astronomers, and which culminated in the great work of Simon Newcomb. A further development of the method of reduction used in this topic, even when refined to take into account the secular variations, is not greatly used in modern times. Because machine and electronic calculation has cut out practically all the drudgery and increased the speed of computation, it is no more difficult to machine process *all* reductions by the rigorous method discussed in the next topic, which has the added advantage of accuracy near the poles. However, for the sake of completeness, the approximate method is illustrated here.

The equations: $$\frac{d\alpha}{dt} = \mu_\alpha + m + n^s \sin\alpha_0 \tan\delta_0 \tag{3.3}$$

$$\frac{d\delta}{dt} = \mu_\delta + n'' \cos\alpha_0 \tag{3.4}$$

where $\frac{d\alpha}{dt}, \frac{d\delta}{dt}$ = the annual rates of change of the coordinates α (seconds of time) and δ (seconds of arc) due to precession and proper motion

μ_α, μ_δ = the annual proper motions in RA (seconds of time) and dec. (seconds of arc) respectively

m = annual precession in RA, defined in Chapter 2, Topic 3, in seconds of time

n^s and n'' = annual precession in dec., defined in Chapter 2, Topic 3, in seconds of time and seconds of arc respectively

α_0 and δ_0 = RA and dec., expressed in decimal degrees, at epoch t_0

If $t - t_0$ is between 5 and 10 years, take m and n for the midpoint of the interval.

Note the similarity with the simplified version of the same equations used in Topic 1 of this chapter, 3.1 and 3.2.

Further information: Among many possible references, the reader is specially directed to W. M. Smart, *Spherical Astronomy*, Chap. X; S. Newcomb, *A Compendium of Spherical Astronomy*, Chap. X; D. McNally, *Positional Astronomy*, Chap. VII.

Example 2: The equatorial coordinates α_0, δ_0, and annual proper motions μ_α, μ_δ, for the star *SAO* 062 191 at epoch 1950.0 are:

$\alpha_0 = 10^h\, 37^m\, 30^s.471 = 159^\circ.376\,963$

$\mu_\alpha = -0^s.032\,2$

$\delta_0 = +31^\circ\, 04'\, 38''.81 = 31^\circ.077\,447\,22$

$\mu_\delta = -0''.087$

The RA and dec. for 1958 are required.

In this case $t - t_0 > 5$ years, so we evaluate m, n^s and n'' for the midpoint of the interval—i.e., 1954—using Eqns. 2.9 and 2.10 for this purpose.

Hence $m = 3^s.073\,34$
$n'' = 20''.042\,2$
$n^s = 1^s.336\,15$

as calculated in Chapter 2, Example 3.

Method 2A

1. Enter δ_0 and evaluate tan	**[31.077 447]** **f tan** **M+**	CS if Southern dec.
2. Enter α_0 and evaluate sin	**[159.376 96]** **f sin**	
3. $\tan\delta_0 \sin\alpha_0$	× **MR**	
4. Multiply by n^s and store	× **[1.336 15]** = **MC** **M+**	
5. Add m	**[3.073 34]** **M+**	
6. Add μ_α	**[0.032 2]** **[CS]** **M+** **MR**	μ_α is negative $\frac{d\alpha}{dt} = 3^s.325$
7. Multiply by number of years	× **[8]**	
8. Note total variation, α	=	$26^s.598$
9. Enter α_0 and evaluate cos	**[159.376 96]** **f cos**	
10. Multiply by n''	× **[20.042 2]**	
11. Add μ_δ	+ **[0.087]** **[CS]** =	μ_δ is negative $\frac{d\delta}{dt} = -18''.84$
12. Multiply by number of years	× **[8]**	
13. Note total variation, δ	=	$-150''.76 = -2'\,30''.76$
14. Add variations to α_0, δ_0		α_0 $10^h\,37^m\,30^s.47$ $+$ $26^s.60$ $\alpha = 10^h\,37^m\,57^s.07$
δ_0 $31°\,04'\,38''.81$		δ_0 $+31°\,04'\,38''.81$ $-$ $02'\,30''.76$ $\delta = +31°\,02'\,08''.05$

Result 2A. The mean coordinates for *SAO* 062 191 at 1958.0 are $\alpha = 10^h\,37^m\,57^s.07$, $\delta = +31°\,02'\,08''.05$. By the rigorous method to be demonstrated later, the mean position at 1958.0 was $\alpha = 10^h\,37^m\,57^s.062$, $\delta = +31°\,02'\,08''.00$. Over this short period of 8 years the error is therefore minimal, $+0^s.01$ in RA, $+0''.05$ in dec., within the claimed limits.

Method 2B

1. Enter: α_0 H.MS	[10.37 30 471]	In H.MS format
	STO 0	
μ_α	[0.032 2]	
	[CHS]	μ_α is negative
	STO 1	
δ_0 D.MS	[31.04 38 81]	CHS if Southern dec.
	STO 2	
μ_δ	[0.087]	
	[CHS]	μ_δ is negative
	STO 3	
2. Enter: m	[3.073 34]	
	STO 4	
n^s	[1.336 15]	
	STO 5	
n''	[20.042 2]	
	STO 6	
Number of years	[8]	
	STO 7	
3. Total variation in RA	RCL 2	
	g → H	
	f tan	
	RCL 0	
	g → H	
	15	
	×	
	STO 0	
	f sin	
	×	
	RCL 5	
	×	
	RCL 4	
	+	
	RCL 1	$(+3^s.32) = \frac{d\alpha}{dt}$
	+	
	RCL 7	
Note variation in RA	×	$+26^s.60$
4. Total variation in dec.	RCL 0	
	f cos	
	RCL 6	
	×	
	RCL 3	$(-18''.84) = \frac{d\delta}{dt}$
	+	
	RCL 7	
Note variation in dec.	×	$-150''.76 = -2'\ 30''.76$
5. Add variations to α_0, δ_0		α_0 $10^h\ 37^m\ 30^s.47$
		$+$ $26^s.60$
		$\alpha = 10^h\ 37^m\ 57^s.07$
		$\delta_0 + 31^\circ\ 04'\ 38''.81$
		$-$ $02'\ 30''.76$
		$\delta = + 31^\circ\ 02'\ 08''.05$
6. For new case go to Step 1, omit Step 2, start again from Step 3		

Result 2B. The mean coordinates for *SAO* 062 191 at 1958.0 are: $\alpha = 10^h 37^m 57^s.07$' $\delta = +31° 02' 08''.05$. By the rigorous method to be demonstrated later, the mean position at 1958.0 was $\alpha = 10^h 37^m 57^s.062$, $\delta = +31° 02' 08''.00$. Over the short period of 8 years the error is therefore minimal, $+0^s.01$ in RA, $+0''.05$ in dec. As written, Method 2B is convenient for a number of similar reductions.

3 Rigorous reduction for the effects of precession and proper motion from the equator and equinox of epoch t_0 to the equator and equinox of a later epoch t (or vice versa), correct to $\pm 0.01^s$ in RA, $\pm 0''.1$ in dec., with a simple 8-digit algebraic calculator (Method 3A); correct to $\pm 0^s.001$ in RA, $\pm 0''.01$ in dec., with a 10-digit RPN calculator (Method 3B).

Introduction: When several similar reductions are to be computed, a great deal of time can be saved by transforming the steps of Method 3B into a standard programme. Not only is time saved, but opportunities for mistakes are reduced to a minimum. If such a facility is provided by the calculator in use, the reader is recommended to use the programmes in Appendix II (Programmes 8 and 9 for the HP-25, Programme 11 for the HP-67 and HP-97).

One should take care not to be misled by the high accuracy achieved over relatively long periods of time when compared with previous methods. Because ζ_0, z and θ are found by means of a series in powers of the time, it must be realized that such a derivation can give reliable results only for a few centuries on either side of the original epoch. Although this cautionary note is necessary, the reader will find that, for reductions over sensible periods, a very high degree of accuracy is obtainable.

The equations:

$$q = \sin\theta\,(\tan\delta_0 + \cos(\alpha_0 + \zeta_0)\tan\tfrac{1}{2}\theta] \qquad (3.5)$$

$$\tan(\Delta\alpha - \mu) = \frac{q\sin(\alpha_0 + \zeta_0)}{1 - q\cos(\alpha_0 + \zeta_0)} \qquad (3.6)$$

$$\mu = \zeta_0 + z \qquad (3.7)$$

$$\alpha = \alpha_0 + \Delta\alpha \qquad (3.8)$$

$$\tan\tfrac{1}{2}(\delta - \delta_0) = \tan\tfrac{1}{2}\theta\,[\cos(\alpha_0 + \zeta_0) - \sin(\alpha_0 + \zeta_0)\tan\tfrac{1}{2}(\Delta\alpha - \mu)] \qquad (3.9)$$

where ζ_0, z and θ are defined in, and evaluated by the method shown in, Chapter 2, Topic 1

α_0 and δ_0 are the equatorial coordinates in RA and dec. at epoch t_0, modified by the effects of proper motion during the time $t - t_0$ (strictly, the equatorial coordinates at epoch t, referred to the equator and equinox of epoch t_0)

α and δ are the equatorial coordinates in RA and dec. at epoch t

Further information: Among many references which could be cited, the reader's attention is especially directed to S. Newcomb, *A Compendium of Spherical Astronomy*, Chap. X; *Explanatory Supplement to the AE*, pp 28–41; Woolard and Clemence, *Spherical Astronomy*, Chap. 13.

Example 3: The equatorial coordinates and annual proper motions of Alpha Ursa Minoris (Polaris) at epoch 1950.0 are given in the *SAO Star Catalogue* as:

(*SAO* 000 308) $\alpha = 1^h\,48^m\,48^s.786$

$\mu_\alpha = +0^s.181\,1$

$\delta = +89°\,01'\,43''.74$

$\mu_\delta = -0''.004$

Reduce to the equator and equinox 1978.0.

Method 3A

1. Evaluate ζ_0, z, $\sin\theta$ and $\tan\frac{1}{2}\theta$ (see Chapter 2, Topic 1)	**MC**	$\zeta_0 = 0.179\,280\,7$ $z = 0.179\,298\,0$ $\sin\theta = 0.002\,720\,6$ $\tan\frac{1}{2}\theta = 0.001\,360\,3$
2. Enter μ_α	**[0.1811]**	CS if μ_α negative
3. Multiply by $t - t_0$	×	
	[28]	CS if reduction is $t \rightarrow t_0$
4. Add seconds of α and convert to minutes	+ **[48.786]** ÷ **60**	
5. Add minutes of α and convert to hours	+ **[48]** ÷ **60**	
6. Add hours of α and convert to degrees	+ **[1]** × **15**	
Note α_0	=	$\alpha_0 = 27°.224\,403$
7. Add ζ_0 (Step 1)	+ **[0.179 280 7]** = **M+**	
Note $\cos(\alpha_0 + \zeta_0)$	**f cos**	0.887 786
	MR	
Note $\sin(\alpha_0 + \zeta_0)$	**f sin**	0.460 257
8. Enter μ_α	**[0.004]**	
	[CS]	μ_α is negative
9. Multiply by $t - t_0$	×	
	[28]	CS if reduction is $t \rightarrow t_0$
10. Add seconds of δ and convert to minutes	+ **[43.74]** ÷ **60**	
11. Add minutes of δ and convert to degrees	+ **[1]** ÷ **60**	
12. Add degrees of δ	+ **[89]**	
Note δ_0	=	$\delta_0 = 89°.028\,785$ CS if Southern dec.
	f tan **MC** **M+**	

Method 3A *continued*

13. Solve for q:		
Enter $\cos(\alpha_0 + \zeta_0)$	**[0.887 786]**	From Step 7
	×	
$\tan\frac{1}{2}\theta$	**[0.001 360 3]**	From Step 1
	+	
$\tan\delta_0$	**MR**	
	×	
$\sin\theta$	**[0.002 720 6]**	From Step 1
	=	($q = 0.160\,483\,8$)
	MC	
	M+	
14. Solve for $\Delta\alpha$:	×	
$\sin(\alpha_0 + \zeta_0)$	**[0.460 257]**	From Step 7
	=	
	←→	
	XM	
	×	
$\cos(\alpha_0 + \zeta_0)$	**[0.887 786]**	From Step 7
	CS	
	+	
	1	
	=	
	←→	
	XM	
	÷	
	MR	
	=	
Note $\Delta\alpha - \mu$	**f tan^{-1}**	$\Delta\alpha - \mu = 4.923\,071$
	MC	
	M+	
ζ_0	**[0.179 280 7]**	
	+	
z	**[0.179 298 0]**	
	+	
	MR	
	=	$\Delta\alpha = 5.281\,649\,7$
15. Add α_0 (Step 6)	+	
	[27.224 403]	
	÷	
	15	
Note hours of α	=	2^h
16. Deduct integral hours and convert to minutes	–	
	[2]	
	×	
	60	
Note minutes of α	=	10^m
17. Deduct minutes and convert to seconds	–	
	[10]	
	×	
	60	
Note seconds of α	=	$01^s.452$
		$\alpha_{1978} = 2^h\,10^m\,01^s.45$

Method 3A *continued*

18. Solve for $\delta - \delta_0$:	**MR**	
	÷	
	2	
	=	
	f tan	
	×	
$\sin(\alpha_0 + \zeta_0)$	**[0.460 257]**	From Step 7
	CS	
	+	
$\cos(\alpha_0 + \zeta_0)$	**[0.887 786**	From Step 7
	×	
$\tan\frac{1}{2}\theta$	**[0.001 360 3]**	From Step 1
	=	
	f $\mathbf{tan^{-1}}$	
	×	
	2	
	=	$\delta - \delta_0 = 0°.135\ 286$
19. Add δ_0 (Step 12)	+	
	[89.028 785]	
Note degrees	=	89°
20. Deduct integral degrees and convert to minutes	−	
	[89]	
	×	
	60	
Note minutes of δ	=	09′
21. Deduct minutes and convert to seconds	−	
	[9]	
	×	
	60	
Note seconds	=	50″.66
		$\delta = 89°\ 09'\ 50''.7$

Result 3A. The mean equatorial coordinates for Polaris for epoch 1978.0 are $\alpha = 2^h\,10^m\,01^s.45$, $\delta = +89°\ 09'\ 50''.7$. The mean place given in the 1978 *AE* is $\alpha = 2^h\,10^m\,01^s.2$, $\delta = +89°\ 09'\ 51''$. Although there is an apparent error of about $+0^s.3$ in RA, there is a reason for this. We took our 1950.0 position from the *SAO Star Catalogue*; the *AE* uses the Boss *General Catalogue* for the base position at 1950.0. Comparison between the *GC* and the *SAO* reveals a difference of $0^s.284$ at 1950.0. (Entries for the *SAO Star Catalogue*, collected from several earlier catalogues—including the *GC*—were first reduced to the system of the *GC*, then to the FK3 system, and finally to the FK4 system.) Our apparent error is thus explained as the result of systematic corrections applied to the *GC* position before incorporation into the *SAO Star Catalogue*.

The example has been set in this manner deliberately, to bring out the point that small differences can occur which stem from the selection of the frame of reference, and not from error in the computation. Now, if you take for the 1950.0 base data, from the *GC*, $\alpha = 1^h\,48^m\,48^s.502$, $\mu_\alpha = +0^s.180\ 7$, $\delta = +89°\ 01'\ 43''.83$, $\mu_\delta = -0''004$, and rework the example, you will obtain, for 1978.0, $\alpha = 2^h\,10^m\,01^s.15$, $\delta = +89°\ 09'\ 50''.7$, which, when rounded, agree with the *AE* data.

The essential point is that, in astrometrical work of the highest accuracy, whenever it would be helpful to other observers the reference system should be quoted.

The accuracy of these reductions for a very close polar star is therefore seen to be excellent, and within the limits claimed.

Method 3B

Step	Keystroke	Notes
1. Evaluate and store ζ_0, z, $\sin\theta$ and $\tan\frac{1}{2}\theta$ (see Chapter 2, Topic 1)		$\zeta_0 = 0°.179\,280\,709$ STO 4 $z = 0°.179\,297\,981$ STO 5 $\sin\theta = 0°.002\,720\,567$ STO 6 $\tan\frac{1}{2}\theta = 0°.001\,360\,286$ STO 7
2. Enter μ_α	**[0.181 1]**	CHS if negative
	↑	'Enter'
3. Multiply by $t - t_0$ and convert to degrees	**[28]**	CHS if reduction is $t \to t_0$
	×	
	3 600	
	÷	
4. Enter α, H.MS format and convert to degrees	**[1.48 48 786]**	
	g → H	
	+	
	15	
	×	
	STO 0	α_0 (including proper motion)
	RCL 4	
	+	
	STO 1	$\alpha_0 + \zeta_0$
5. Enter μ_δ	**[0.004]**	
	[CHS]	μ_δ is negative
	↑	'Enter'
6. Multiply by $t - t_0$ and convert to degrees	**[28]**	CHS if reduction is $t \to t_0$
	×	
	3 600	
	÷	
7. Enter δ, D.MS	**[89.01 43 74]**	CHS if Southern dec.
	g → H	
	+	
	STO 2	δ_0 (including proper motion)
8. Solve for q:	**f tan**	
	RCL 7	
	RCL 1	
	f cos	
	×	
	+	
	RCL 6	
	×	
	STO 3	($q = 0.160\,484\,912$)
9. Solve for $\Delta\alpha$:	**RCL 1**	
	f sin	
	×	
	RCL 3	
	RCL 1	
	f cos	
	×	
	CHS	
	1	
	+	

Method 3B *continued*

	÷	
	g $\tan^{-1}$	
	STO 3	($\Delta\alpha - \mu = 4.923\ 118\ 628$)
		q is no longer required
	RCL 4	
	RCL 5	
	+	
	+	($\Delta\alpha = 5.281\ 697\ 318$)
10. Add for α	**RCL 0**	
	+	
	15	
	÷	
Note integral hours		2^h
	g FRAC	
	60	
Note minutes	×	10^m
	g FRAC	
	60	
Note seconds	×	$01^s.464$
11. Solve for δ:	**RCL 3**	
	2	
	÷	
	f tan	
	RCL 1	
	f sin	
	×	
	CHS	
	RCL 1	
	f cos	
	+	
	RCL 7	
	×	
	g $\tan^{-1}$	
	2	
	×	
	RCL 2	
Note degrees	+	89°
	g FRAC	
	60	
Note minutes	×	09′
	g FRAC	
	60	
Note seconds	×	50″.713
		$\alpha = 2^h\ 10^m\ 01^s.464$
		$\delta = +89°\ 09'\ 50''.71$

Result 3B. The mean equatorial coordinates for Polaris for epoch 1978.0 are $\alpha = 2^h\ 10^m\ 01^s.464$, $\delta = +89°\ 09'\ 50''.71$. The *AE* for 1978 gives $\alpha = 2^h\ 10^m\ 01^s.2$, $\delta = +89°\ 09'\ 51''$. See the explanation with the result of Method 3A regarding the systematic corrections applied to the *GC* coordinates before incorporation in the *SAO Star Catalogue*.

The accuracy of the result obtained with the 10-digit RPN calculator is excellent.

For further practice, try the following:

(a) Reduce the equatorial coordinates α, δ for the following stars from epoch 1950.0 to 1977.0, to the nearest $0^s.1$ in RA, to the nearest second in dec. Use Topic 1 of Chapter 2 to derive the precessional constants.

		α	μ_α	δ	μ_δ
(*i*)	θ And	$0^h\ 14^m\ 28^s.299$	$-0^s.004\ 4$	$+38°\ 24'\ 14''.90$	$-0''.015$
(*ii*)	β Ret	$3^h\ 43^m\ 33^s.963$	$+0^s.049\ 5$	$-64°\ 57'\ 50''.21$	$+0''.078$
(*iii*)	α Cam	$4^h\ 49^m\ 03^s.825$	$+0^s.000\ 6$	$+66°\ 15'\ 38''.64$	$+0''.008$
(*iv*)	β UMi	$14^h\ 50^m\ 49^s.645$	$-0^s.008\ 6$	$+74°\ 21'\ 35''.58$	$+0''.010$
(*v*)	β Oct	$22^h\ 41^m\ 04^s.412$	$-0^s.028\ 4$	$-81°\ 38'\ 41''.05$	$+0''.006$

(b) Using the methods illustrated in Chapter 2, Topic 1, compute the precessional constants necessary to reduce the mean equatorial coordinates of Polaris for 1978.0 to the equator and equinox of 1755.0 and then carry out the reduction, to the nearest $0^s.1$ in RA, to $0''.1$ in dec., given $\alpha_{1978} = 2^h\ 10^m\ 01^s.46$, $\mu_\alpha = +0^s.208\ 1$, $\delta_{1978} = +89°\ 09'\ 50''.71$, $\mu_\delta = -0''.008$.

(c) Auwers' catalogue of Bradley stars, epoch 1755.0, gives the position of Beta Ursa Minoris as $\alpha = 14^h\ 51^m\ 42^s.56$, $\delta = +75°\ 09'\ 23''.2$. Ignoring the effect of proper motion (input 0 in the computation), reduce these coordinates to epoch 1875.0, an interval of 120 years.

(d) *Apparent Places of Fundamental Stars*, 1977, quotes the mean place for Epsilon Cassiopeiae at 1977.0 as $1^h\ 52^m\ 43^s.380$, $+63°\ 33'\ 27''.54$. Given, from the *SAO Star Catalogue*, the 1950.0 coordinates and proper motions: $\alpha = 1^h\ 50^m\ 46^s.378$, $\mu_\alpha = +0^s.004\ 9$, $\delta = +63°\ 25'\ 29''.89$, $\mu_\delta = -0''.015$; and using Method 3B with the precessional constants derived from Topic 1 of Chapter 2, how does your result compare with the mean place for 1977.0 quoted above?

Your answers should be:

(a)	(*i*)	$0^h\ 15^m\ 53^s.1$	$+38°\ 33'\ 14''$
	(*ii*)	$3^h\ 43^m\ 54^s.4$	$-64°\ 52'\ 45''$
	(*iii*)	$4^h\ 51^m\ 45^s.2$	$+66°\ 18'\ 21''$
	(*iv*)	$14^h\ 50^m\ 45^s.3$	$+74°\ 14'\ 58''$
	(*v*)	$22^h\ 43^m\ 47^s.5$	$-81°\ 30'\ 11''$

If you check by confirming with the 1977 *AE*, you will note that, although the declinations agree with the ephemeris, there were some small deviations in the 1977 right ascensions. This is because the compilers of the *AE* used as base for 1950.0 the coordinates listed in the Boss *General Catalogue*, while in the question I used those listed in the *SAO Star Catalogue*. See the note on this point appended to Result 3A. If the *GC* coordinates and proper motions are used (or those in the *Atlas Cæli Catalogue*) the results agree exactly with the mean positions for 1977.0 shown in the *AE*.

(b) The 1755 coordinates are $\alpha = 0^h\ 43^m\ 41^s.9$, $\delta = +87°\ 59'\ 41''.4$. If you did not obtain this result, check that you remembered to use $\zeta_0 = -z$, $z = -\zeta_0$, and that you changed the sign of θ, $\tan\frac{1}{2}\theta$ and $\sin\theta$. Also, when computing the total proper motions over the period, that you changed the sign when entering the number of years, to –223.

(c) This example is worked in §144, *A Compendium of Spherical Astronomy*, where the result is given as $\alpha = 14^h\,51^m\,06^s.35$, $\delta = +74°\,39'\,58''.82$. Your answer should be within $\pm 0^s.01$ in RA, $\pm 0''.1$ in dec.
(d) The result from Method 3B is $\alpha = 1^h\,52^m\,43^s.381$, $\delta = +63°\,33'\,27''.54$ which, compared with the coordinates given by *Apparent Places*, is in error by $+0^s.001$ for α, and in exact agreement for δ. The claimed accuracy is achieved.

4 Rotational geometry and an alternative method of rigorous reduction for precession and proper motion.

Introduction: Printed catalogues normally list stars in terms of the equatorial coordinates α, δ in ascending order of α (although sometimes divided into convenient bands or zones of declination, e.g., *BD* and *SAO*). However, there are alternative methods nowadays of storing and retrieving this information (punched cards, magnetic tape, disc, etc.) where it might be just as convenient for data-processing purposes to record the positions in terms of the rectangular equatorial coordinates, x, y, z. By three successive rotations about the three rectangular axes, z_0, y', z^*, through the angles ζ_0, θ and z respectively, it is possible to transform the initial reference frame to the equator and equinox of another epoch.

Further information: The method is fully developed in D. McNally, *Positional Astronomy*, 7.2, and the reader is encouraged to refer to this treatment for further details. An even more complete treatment of matrix algebra and its application to transformation of coordinates is found in I. I. Mueller, *Spherical and Practical Astronomy as applied to Geodesy*, 3.34, 3.35, 4.333, 4.422 and 4.424.

Programme: Most Hewlett-Packard calculators have a powerful feature: polar magnitude and angle conversion into rectangular x, y coordinates, and vice versa, for vector work. This feature can be employed to perform the three rotations referred to in the introductory paragraph. The method is really suitable only for computer techniques or powerful programmable calculators such as those mentioned.

Readers having access to such a calculator will find it instructive to use the specially-written programmes in the Appendix (Programmes 12 and 13), and to re-work the 'further practice' problems of Topic 3, using the rotational-geometry method as an alternative to the trigonometric methods illustrated there.

Accuracy is identical to that achieved by Method 3B and its equivalent programme in the Appendix. The lengths of the alternative programmes are about equal, so there is no saving of time in entering or in subsequent key-strokes which might otherwise give one programme the edge over the other. It is simply a matter of preference whether the reader elects to work in terms of α, δ or x, y, z coordinates.

As written, the alternative programme incorporates initial conversion from α, δ to x, y, z coordinates, and vice versa at the end of the transformation, for convenience in working from conventional printed star catalogues. If the base material is already available in terms of x, y, z coordinates then these conversion routines can be deleted or bypassed; this will shorten the programme entry time (for the HP-25 and HP-55, which do not work from magnetic cards as do the other two in the family) but it will be found that the running time will not be reduced significantly.

NOTES

4 Reduction from Mean to Apparent Place

Topic Reduction from mean place at the start of a Besselian solar year to apparent place at any time during the year, with allowance for precession, nutation, aberration and proper motion.

1 To reduce the equatorial coordinates of a star, α, δ, from the mean place to the apparent place of date, correct to $\pm 0^s.001$ in RA, to $\pm 0''.01$ in dec.

Introduction: The methods of computation demonstrated in Chapter 3 enable the coordinates of a star, referred to the equator and equinox of a standard catalogue epoch, to be reduced to the equator and equinox of any other epoch (the start of a Besselian solar year), with allowance for the effects of precession and proper motion. The mean coordinates are correct for that instant near the start of the selected calendar year. (See Chapter 9, Topic 1 for the method of calculating the beginning of the Besselian solar year.)

When revised coordinates are required for times after the epoch a further reduction is necessary, to give the apparent position of date. The corrections include allowance for precession, nutation, aberration and proper motion. If a final correction for parallax is also desired (and this will only rarely be necessary) this may be computed separately.

The corrections are determined by the use of the Besselian Day Numbers, which are tabulated daily in the *AE*. For the purposes of the computation it is assumed the reader will have access to the *AE* for the required year. Where this is not possible—e.g., before publication of the relevant *AE*—the computer will have to calculate the values of the Day Numbers himself; the computation for *A*, *B* and *E* is straightforward, but that for *C*, *D*, *J* and *J'* is more tedious. The subject is dealt with in detail in Chapter 5D of the *Explanatory Supplement to the AE*. Topic 1 of Chapter 9 of this book will enable reasonably accurate approximations to be derived in the minimum of time.

An HP-67 programme for reduction to apparent place is included in the Appendix; it will be found especially useful for occultation work (Programme 56).

The equations:

$$\alpha = \alpha_0 + \tau\mu_\alpha + Aa + Bb + Cc + Dd + E + J\tan^2\delta_0 \quad (4.1)$$
$$\delta = \delta_0 + \tau\mu_\delta + Aa' + Bb' + Cc' + Dd' + J'\tan\delta_0 \quad (4.2)$$

where zero subscripts denote the mean place at the start of the year

$\tau = \dfrac{t}{365.242\ 2}$ where t is the number of days from the *nearest* beginning of a Besselian solar year

A, B, C, D, E, J, J' are the Besselian Day Numbers

$a, b, c, d, a', b', c', d'$ are Besselian Star Constants, where:

$a = \dfrac{m}{n} + \sin\alpha_0 \tan\delta_0$

$b = \cos\alpha_0 \tan\delta_0$

$c = \cos\alpha_0 \sec\delta_0$

$d = \sin\alpha_0 \sec\delta_0$

$a' = \cos\alpha_0$

$b' = -\sin\alpha_0$

$c' = \tan\epsilon \cos\delta_0 - \sin\alpha_0 \sin\delta_0$

$d' = \cos\alpha_0 \sin\delta_0$

and m, n and ϵ are as defined in Chapter 2.

Further information: *Explanatory Supplement to the AE*, Chap. 5; any current *AE*, on or near p 547; Woolard and Clemence, *Spherical Astronomy*, p 283 *et seq*; D. McNally, *Positional Astronomy*, pp 175–179; W. M. Smart, *Spherical Astronomy*, pp 242–246.

Example: Using the mean place for Epsilon Cassiopeiae at epoch 1978.0, (which will be found by the method of Topic 3 of Chapter 3 to be $\alpha = 1^h\,52^m\,47^s.729$, $\delta = +63°\,33'\,45''.19$) and the proper motion (which by Chapter 5 will be found to be unchanged from its value at 1950.0: $\mu_\alpha = +0^s.0049$, $\mu_\delta = -0''.015$), compute to the second order the apparent place at Greenwich upper transit on 1977, November 11.937, given the following Besselian Day Numbers:

0^h ET	τ	A	B	C	D	E
Nov 11	−0.138 8	−1″.815	+9″.125	+12″.468	+15″.344	+0ˢ.000 4
Nov 12	−0.136 1	−1″.745	+9″.185	+12″.218	+15″.582	+0ˢ.000 4

	J	J'
Nov 11	0	−0″.001 6
Nov 12	+0ˢ.000 01	−0″.001 6

and

	m	n	ϵ
1977.0	46″.106 6	20″.040 3	23°.442 274
1978.0	46″.106 9	20″.040 2	23°.442 144

Before the apparent place for upper transit can be computed, values for the Day Numbers must be interpolated to November 11.937, and values for m, n and ϵ interpolated to 1977 November, assuming the change in rates to be linear over these periods. Then, if x is the value of the required Day Number, y its value on November 11 at 0^h ET and z its value on November 12, in this case: $x = z-(1-0.937)(z-y)$.

Thus, we find for November 11.937:
$\tau = -0.136\,3$, $A = -1''.749$, $B = +9''.181$, $C = +12''.234$, $D = +15''.567$, $E = +0^s.000\,4$, $J = +0^s.000\,01$, $J' = -0''.001\,6$.

By similar interpolation technique we find:
$m = 46''.106\,9$, $n = 20''.040\,2$, $\epsilon = 23°.442\,162$.

The foregoing interpretation presupposes that the time of Greenwich upper transit is known on the required date. But what if the transit time is not known beforehand? An interpolation on a different basis can be made in these circumstances, as the *AE* lists (in the next following section) the Besselian Day Numbers, A, B, C, D, for 0^h ST daily. Suppose, then, that the Greenwich upper transit time for Epsilon Cassiopeiae on 1977, November 11 is unknown. In this case, take the Day Numbers for 0^h ST on November 11 and 12, and interpolate to the right ascension of the star:

0^h ST	A	B	C	D
Nov 11	$-1''.756$	$+9''.177$	$+12''.254$	$+15''.549$
Nov 12	$-1''.672$	$+9''.227$	$+12''.000$	$+15''.781$

and take E, J and J' as before.

Interpolation is to $1^h\,52^m\,47^s.729 = 1^h.879\,9$ and, putting x as the required Day Number, y its value on Nov 11 at 0^h ST, z its value on Nov 12:

$$x = y + \frac{\alpha\,(z-y)}{24}.$$

Thus, we obtain for the Greenwich upper transit on 1977, November 11, $A = -1''.749$, $B = +9''.181$, $C = +12''.234$, $D = +15''.567$, which values are seen to be identical to those previously obtained by interpolation when the transit time was known.

Method A

1. Find $A\alpha$:		
Enter α_0, in degrees	**[28.198 871]**	
	f sin	
	M+	
Enter δ_0, in degrees	**[63.562 553]**	CS if Southern dec.
	f tan	
	$\times$	
	MR	
	=	
	MC	
	M+	
Enter m	**[46.106 9]**	
	$\div$	
Enter n	**[20.040 2]**	

Method A *continued*

	=	
	M+	
Enter A	[1.749]	
	[CS]	A is negative
	×	
	MR	
	÷	
	15	
Note Aa	=	($Aa = -0^s.379\,07$)
2. Find Bb:		
Enter α_0	[28.198 871]	
	f cos	
	MC	
	M+	
Enter δ_0	[63.562 553]	CS if Southern dec.
	f tan	
	×	
	MR	
	×	
Enter B	[9.181]	
	÷	
	15	
Note Bb	=	($Bb = +1^s.084\,88$)
3. Find Cc:		
Enter α_0	[28.198 871]	
	f cos	
	MC	
	M+	
Enter δ_0	[63.562 553]	
	f cos	
	f $\frac{1}{x}$	
	×	
	MR	
	×	
Enter C	[12.234]	
	÷	
	15	
Note Cc	=	($Cc = +1^s.614\,48$)
4. Find Dd:		
Enter α_0	[28.198 871]	
	f sin	
	MC	
	M+	
Enter δ_0	[63.562 553]	CS if Southern dec.
	f cos	
	f $\frac{1}{x}$	
	×	
	MR	
	×	
Enter D	[15.567]	
	÷	
	15	
Note Dd	=	($Dd = +1^s.101\,46$)

Method A *continued*

5. Find $J \tan^2\delta_0$:		
Enter δ_0	**[63.562 553]**	CS if Southern dec.
	f tan	
	×	
	=	
	×	
Enter J in seconds	**[0.000 01]**	
Note $J \tan^2\delta_0$	=	($J \tan^2\delta_0 = +0^s.000\ 04$)
6. Find $\tau\mu_\alpha$:		
Enter μ_α in seconds	**[0.004 9]**	CS if μ_α is negative
	×	
Enter τ	**[0.136 3]**	
	[CS]	τ is negative
Note $\tau\mu_\alpha$	=	($\tau\mu_\alpha = -0^s.000\ 67$)
7. Find $\Delta\alpha$ in seconds:		
Enter Aa, Step 1	**[0.379 07]**	
	[CS]	Aa is negative
	+	
Enter Bb, Step 2	**[1.084 88]**	
	+	
Enter Cc, Step 3	**[1.614 48]**	
	+	
Enter Dd, Step 4	**[1.101 46]**	
	+	
Enter E in seconds	**[0.000 4]**	
	+	
Enter $J \tan^2\delta_0$, Step 5	**[0.000 04]**	
	+	
Enter $\tau\mu_\alpha$, Step 6	**[0.000 67]**	
	[CS]	$\tau\mu_\alpha$ is negative
	+	
Enter seconds of α_0	**[47.729]**	
Read $\Delta\alpha$ in seconds	=	($51^s.151$)
If display is negative, add 60 seconds, note new seconds, and reduce minutes of α_0 by 1		$\alpha = 1^h\ 52^m\ 51^s.151$
8. Find Aa':		
Enter α_0	**[28.198 871]**	
	MC	
	M+	
	f cos	
	×	
Enter A	**[1.749]**	
	[CS]	A is negative
Note Aa'	=	($Aa' = -1''.541\ 4$)
9. Find Bb':	**MR**	
	f sin	
	CS	
	×	
Enter B	**[9.181]**	
Note Bb'	=	($Bb' = -4''.338\ 3$)
10. Find Cc':		
Enter ϵ	**[23.442 162]**	
	f tan	

Method A *continued*

	MC	
	M+	
Enter δ_0	**[63.562 553]**	CS if Southern dec.
	f cos	
	×	
	MR	
Note first term	=	(0.193 053 1)
Enter α_0	**[28.198 871]**	
	f sin	
	MC	
	M+	
Enter δ_0	**[63.562 553]**	CS if Southern dec.
	f sin	
	×	
	MR	
	=	
	MC	
	M+	
Enter first term	**[0.193 053 1]**	
	−	
	MR	
	×	
Enter *C*	**[12.234]**	
Note *Cc′*	=	($Cc' = -2''.814\,6$)
11. Find *Dd′*:		
Enter α_0	**[28.198 871]**	
	f cos	
	MC	
	M+	
Enter δ_0	**[63.562 553]**	
	f sin	
	×	
	MR	
	×	
Enter *D*	**[15.567]**	
Note *Dd′*	=	($Dd' = +12''.284\,6$)
12. Find $J' \tan\delta_0$:		
Enter δ_0	**[63.562 553]**	
	f tan	
	×	
Enter *J′*	**[0.001 6]**	
	[CS]	*J′* is negative
Note $J' \tan\delta_0$	=	($J' \tan\delta_0 = -0''.003\,2$)
13. Find $\tau\mu_\delta$:		
Enter μ_δ	**[0.015]**	
	[CS]	μ_δ is negative
	×	
Enter τ	**[0.136 3]**	
	[CS]	τ is negative
Note $\tau\mu_\delta$	=	($\tau\mu_\delta = +0''.002\,0$)
14. Find $\Delta\delta$ in arcsecs:	+	
Aa′, Step 8	**[1.541 4]**	
	[CS]	*Aa′* is negative
	+	

Method A *continued*

Bb', Step 9	[4.338 3]	
	[CS]	Bb' is negative
	+	
Cc', Step 10	[2.814 6]	
	[CS]	Cc' is negative
	+	
Dd', Step 11	[12.284 6]	
	+	
$J' \tan\delta_0$, Step 12	[0.003 2]	
	[CS]	$J' \tan\delta_0$ is negative
	+	
Add seconds of δ_0	[45.19]	
Read $\Delta\delta$ in seconds	=	$\Delta\delta = 48''.779$
If display is negative, add 60 seconds and reduce minutes of δ_0 by 1		$\delta = +63° 33' 48''.78$

Result A. The coordinates of Epsilon Cassiopeiae at Greenwich upper transit on 1977, November 11 were $\alpha = 1^h 52^m 51^s.151$, $\delta = +63° 33' 48''.78$. This position includes allowance for the short-period terms of nutation, but not for parallax. (See the additional note with the result for Method B for comparison with the position given by *Apparent Places of Fundamental Stars*, 1977.)

Method B

1. Find $\Delta\alpha$:		
Enter α_0 in H.MS format	[1.52 47 729]	
	g → H	
	15	
	×	
	STO 0	
Enter δ_0 in D.MS format	[63.33 45 19]	CHS if Southern dec.
	g → H	
	STO 1	
Enter m in arcsecs	[46.106 9]	
	↑	'Enter'
Enter n''	[20.040 2]	
	÷	
	RCL 0	
	f sin	
	RCL 1	
	f tan	
	×	
	+	
Enter A	[1.749]	
	[CHS]	A is negative
	STO 2	
	×	
	15	
	÷	
	STO 7	
	RCL 0	
	f cos	

Method B *continued*

	RCL 1	
	f tan	
	×	
Enter *B*	**[9.181]**	
	STO 3	
	×	
	15	
	÷	
	STO + 7	
	RCL 0	
	f cos	
	RCL 1	
	f cos	
	g $\frac{1}{x}$	
	×	
Enter *C*	**[12.234]**	
	STO 4	
	×	
	15	
	÷	
	STO + 7	
	RCL 0	
	f sin	
	RCL 1	
	f cos	
	g $\frac{1}{x}$	
	×	
Enter *D*	**[15.567]**	
	STO 5	
	×	
	15	
	÷	
	STO + 7	
Enter *E*	**[0.000 4]**	
	STO + 7	
Enter *J*	**[0.000 01]**	
	RCL 1	
	f tan	
	g x^2	
	×	
	STO + 7	
Enter μ_a	**[0.004 9]**	
	↑	'Enter'
Enter τ	**[0.136 3]**	
	[CHS]	τ is negative
	STO 6	
	×	
	STO + 7	
Enter seconds of a_0	**[47.729]**	
	RCL 7	
	+	

Method B *continued*

Note new seconds of α; if display is negative, add 60 seconds and reduce minutes of α_0 by 1		$(51^s.151)$ $\alpha = 1^h\ 52^m\ 51^s.151$
2. Find $\Delta\delta$:	**RCL 0**	
	f cos	
	RCL 2	
	×	
	STO 7	
	RCL 0	
	f sin	
	CHS	
	RCL 3	
	×	
	STO + 7	
Enter ϵ	**[23.442 162]**	
	f tan	
	RCL 1	
	f cos	
	×	
	RCL 0	
	f sin	
	RCL 1	
	f sin	
	×	
	−	
	RCL 4	
	×	
	STO + 7	
	RCL 0	
	f cos	
	RCL 1	
	f sin	
	×	
	RCL 5	
	×	
	STO + 7	
	RCL 1	
	f tan	
Enter J'	**[0.001 6]**	
	[CHS]	J' is negative
	×	
	STO + 7	
Enter μ_δ	**[0.015]**	
	[CHS]	μ_δ is negative
	RCL 6	
	×	
	STO + 7	
Enter seconds of δ_0	**[45.19]**	
	RCL 7	
	+	
Note new seconds of δ; if display is negative, add 60 seconds and reduce minutes of δ_0 by 1		$(48''.78)$ $\delta = +63^\circ\ 33'\ 48''.78$

Result B. The coordinates of Epsilon Cassiopeiae at Greenwich upper transit on 1977, November 11 were $\alpha = 1^h\,52^m\,51^s.151$, $\delta = +63°\,33'\,48''.78$. This position includes allowance for the short-period terms of nutation, but not parallax.

The position given by *Apparent Places of Fundamental Stars*, 1977, for this transit is $\alpha = 1^h\,52^m\,51^s.160$, $\delta = +63°\,33'\,48''.86$, but this excludes the short-period terms because of interpolation difficulties.

As a check we can compute the short-period terms separately and add them to the position given in *Apparent Places*.

The values for f', g' and G' are given in the *AE*:

	f'	g'	G'
Nov 11	$-0^s.0130$	$0''.085$	$11^h\,49^m$
Nov 12	$-0^s.0103$	$0''.079$	$9^h\,52^m$

Interpolating, as before, to November 11.937 we obtain:
$f' = -0^s.010\,5$, $g' = 0''079\ (0^s.005\,3)$, $G' = 9^h\,59^m$.

Then, $\Delta\alpha = f' + g' \sin(G' + \alpha_0) \tan\delta_0$

$\Delta\delta = g' \cos(G' + \alpha_0)$

$\Delta\alpha = -0^s.010$

$\Delta\delta = -0''.079$

Adding these terms to the coordinates given by *Apparent Places* gives the required position, including short-period terms: $\alpha = 1^h\,52^m\,51^s.150$, $\delta = +63°\,33'\,48''.78$, which agrees to within $\pm 0^s.001$ in RA, and exactly in dec., with the result obtained by Methods A and B. Thus the accuracy of the method is excellent and needs only the additional corrections, if required, for parallax and refraction.

The worked examples in this Chapter show the complete method of reduction to the second order. If the nature of the work (e.g., apparent places of stars for occultation work) demands accuracy to the first order only, the terms in Eqns. 4.1 and 4.2 which include J and J' can be dropped. In this event, HP-67 and HP-97 users should employ Programme 56 in the Appendix.

NOTES

5 Proper Motion

Topic To calculate the effect of precession on proper motion.

1 To calculate the change in proper motion, μ_α, μ_δ, with precession.

Introduction: μ_α, μ_δ, being the components of proper motion μ, φ in RA and dec. respectively, must clearly change with the times because φ is related to the North point of a particular epoch. This secular change is most obvious near the poles.

With the exception of a few nearby stars which have exceptionally large proper motion, so that a factor of acceleration has to be taken into account, μ is a constant over long periods of time. On the other hand, φ will be changing due to the movement of the North Celestial Pole (NCP).

In this topic, we shall consider various methods of reducing μ_α, μ_δ from one epoch to another.

Further information: The theoretical aspects are treated in detail in W. Chauvenet, *A Manual of Spherical and Practical Astronomy*, Vol. 1, pp 621–623; S. Newcomb, *A Compendium of Spherical Astronomy*, pp 260–265; D. Smart, *Spherical Astronomy*, Chap. XI; D. McNally, *Positional Astronomy*, pp 180–9.

Method 1.

The simplest method of determining μ_α, μ_δ at any epoch, when the proper motion is known (from a catalogue) for some particular epoch, can be employed when reducing the mean position of the star from that of the catalogue to the same new epoch. In Chapter 3, Example 3, the known equatorial coordinates of Alpha Ursa Minoris (Polaris) were reduced from 1950.0 to the equator and equinox of 1978.0, using for μ_α and μ_δ the values given in the *SAO Star Catalogue* for 1950.0: $\mu_\alpha = +0^s.181\,1$, $\mu_\delta = -0''.004$. The 1978.0 coordinates thus reduced were: $\alpha = 2^h\,10^m\,01^s.464$, $\delta = +89°\,09'\,50''.71$.

Suppose, now, the reduction of position is carried out again, this time ignoring the proper motions (i.e., input 0 at the appropriate steps). In this case, Method 3B of Chapter 3 gives the 1978.0 coordinates: $\alpha = 2^h\,09^m\,55^s.638$, $\delta = +89°\,09'\,50''.94$.

The difference between these two sets of coordinates is due to the proper motion of the star during the interval between the epochs. If this difference is divided by the number of years in the period, the result is the annual proper motion, in RA and dec. respectively, at the new epoch of 1978.0:

$5^s.826 \div 28 = +0^s.208\ 1$, $-0''.230 \div 28 = -0''.008$.

Thus, the proper motions at 1978.0 are $\mu_\alpha = +0^s.208\ 1$, $\mu_\delta = -0''.008$.

Method 2.

Alternatively, once the reduction for mean place from the known epoch to the required new epoch has been carried out, the revised declination can be used in Chauvenet's equations:

$$\sin\gamma = \frac{\sin\theta \sin(\alpha_0 + \zeta_0)}{\cos\delta'} \tag{5.1}$$

$$\cos\delta' \times \mu_\alpha' = \mu_{\alpha 0} \cos\delta_0 \cos\gamma + \frac{\mu\delta_0}{15} \sin\gamma \tag{5.2}$$

$$\mu_\delta' = -15\mu_{\alpha 0} \cos\delta_0 \sin\gamma + \mu_{\delta 0} \cos\gamma \tag{5.3}$$

where θ and ζ_0 are the same as used in the main reduction, μ_α' and $\mu_{\alpha 0}$ are expressed in seconds of time, μ_δ' and $\mu_{\delta 0}$ are expressed in seconds of arc, superscript $'$ denotes the new epoch and subscript $_0$ the old epoch.

Taking the same example as in Method 1, that is, the reduction of Polaris from 1950.0 to 1978.0, and solving Eqn. 5.1 for γ:

$$\sin\gamma = \frac{0.002\ 720\ 567 \sin(27^\circ.203\ 275\ 01 + 0^\circ.179\ 280\ 709)}{\cos 89^\circ.164\ 086\ 11}$$

$$= \frac{0.001\ 251\ 269}{0.014\ 588\ 933}$$

$$= 0.085\ 768\ 370.$$

$\therefore \gamma = 4^\circ.920\ 210\ 345$ and so

$\cos\gamma = 0.996\ 315\ 104.$

Then solve for μ_α' by Eqn. 5.2:

$$0.014\ 588\ 933\mu_\alpha' = (0.181\ 1 \times 0.016\ 949\ 535 \times 0.996\ 315\ 104) + \left(\frac{-0.004}{15} \times 0.085\ 768\ 370\right)$$

$$= 0^s.003\ 035\ 378.$$

$$\mu_\alpha' = \frac{0.003\ 035\ 378}{0.014\ 588\ 933}$$

$$= +0^s.208\ 06.$$

Lastly, solve for μ_δ' by Eqn. 5.3:

$$\mu_\delta' = (-15 \times 0.181\ 1 \times 0.016\ 949\ 535 \times 0.085\ 768\ 370) + (-0.004 \times 0.996\ 315\ 104)$$

$$= -0.003\ 949 + (-0.003\ 985)$$

$$= -0''.007\ 9.$$

Thus, the proper motions of Polaris at 1978.0 are $\mu_\alpha = +0^s.208\ 1$, $\mu_\delta = -0''008$, which result agrees with that of Method 1.

Method 3.

Woolard and Clemence, in *Spherical Astronomy*, give the rigorous equations:

$$\mu_\alpha' = \mu_{\alpha 0}[\cos\theta + \sin\theta \tan\delta' \cos(\alpha' - z)] + \frac{1/15\ \mu\delta_0 \sin\theta \sin(\alpha' - z)}{\cos\delta_0 \cos\delta'} \tag{5.4}$$

$$\mu\delta' = -15\ \mu_{\alpha 0} \sin\theta \sin(\alpha' - z) + \mu_{\delta 0}[\cos\theta + \sin\theta \tan\delta' \cos(\alpha' - z)] \times \frac{\cos\delta'}{\cos\delta_0} \tag{5.5}$$

This reduces the number of equations to two, but each is of great length. The use of these equations leads to very accurate results, but to the four decimal places normally used for μ_α and the three for μ_δ they give the same results as those of Chauvenet, even for close polar stars. Method 2 will therefore generally be found to be more convenient from a practical point of view.

Method 4.

A short cut might possibly be employed. Note, in Step 14 of Method 3A and Step 9 of Method 3B of Chapter 3, that $\Delta\alpha - \mu$ approximates the value of γ in Eqn. 5.1. For example, if the value of $\Delta\alpha - \mu$ for Polaris given by Method 3B, Chapter 3, (4°.923 118 628) is employed also for γ, thus eliminating Eqn. 5.1, and μ_α' and μ_δ' are solved by Eqns. 5.2 and 5.3 using this approximation, the results are:

$$\mu_\alpha = +0^s.208\,1, \; \mu_\delta = -0''.008,$$

agreeing with Methods 1 and 2.

Even over longer intervals—e.g., 1900 → 1978—the use of $\Delta\alpha - \mu$ from the reduction will give accurate results. Therefore, when carrying out reductions for mean place from one epoch to another by the rigorous method of Topic 3, Chapter 3, note the value of $\Delta\alpha - \mu$ for use as γ in Eqns. 5.2 and 5.3 if it is desired to revise the annual proper motions to the new epoch. If using Method 3B, there is no need especially to note down $\Delta\alpha - \mu$ at Step 9—it can be retrieved at the end of the computation by 'RCL 3'. A sub-routine for the HP-25 programmes in the Appendix has been devised for use when a number of such reductions is undertaken, giving the proper motions at the new epoch automatically (Programme 10).

For further practice, try the following:

(a) The *SAO Star Catalogue* gives the equatorial coordinates and annual proper motions for Epsilon Cassiopeiae at 1950.0 as:

$$\alpha = 1^h\,50^m\,46^s.378,$$
$$\mu_\alpha = +0^s.004\,9,$$
$$\delta = +63^\circ\,25'\,29''.89,$$
$$\mu_\delta = -0''.015.$$

Using the method of Topic 3, Chapter 3, the mean coordinates for 1978.0 will be found to be: $\alpha = 1^h\,52^m\,47^s.729$, $\delta = +63^\circ\,33'\,45''.19$. Re-work the example to find $\Delta\alpha - \mu$ and, by Method 4 above, find the annual proper motions for 1978.0.

(b) The mean place for Polaris at 1978.0 was found in Topic 3 of Chapter 3 to be: $\alpha = 2^h\,10^m\,01^s.464$, $\delta = +89^\circ\,09'\,50''.71$ in the FK4 system. Methods 1 and 2 of this chapter give the annual proper motions at 1978.0 as: $\mu_\alpha = +0^s.208\,1$, $\mu_\delta = -0''.008$. Find the mean position and proper motion for 1900.0. Use Topic 1 of Chapter 2 to derive the precessional constants.

Your answers should be:

(a) $\Delta\alpha - \mu = 0^\circ.146\,477\,378$; the proper motions are unchanged from 1950.0.

(b) The mean place of Polaris at 1900.0 was $\alpha = 1^h\,22^m\,33^s.645$, $\delta = +88^\circ\,46'\,26''.46$, and the proper motion $\mu_\alpha = +0^s.144\,0$, $\mu_\delta = +0''.001$.

The value of the 1900.0 proper motion can be checked by reducing the 1950.0 coordinates to 1900.0, ignoring proper motion, deducting from the result the

1900.0 coordinates previously derived, and dividing by 50. The precessional constants for 1950 → 1900 are $\zeta_0 = -0.320\,111\,817$, $z = -0.320\,056\,757$, $\tan\frac{1}{2}\theta = -0.002\,429\,480$, $\sin\theta = -0.004\,858\,932$. The proper motion thus found for 1900.0 is again given as $\mu_\alpha = +0^s.144\,0$, $\mu_\delta = +0''.001$.

Determination of μ, φ.

In the introduction it was noted that μ is effectively constant over long intervals of time, the only exceptions being a very few stars close to the Sun which have exceptionally large proper motions. (McNally comments on Barnard's Star, and Newcomb on 1830 Groombridge, for which the perspective accelerations are $+0''.001\,2$ annually and $+0''.000\,19$ respectively.)

Because in practice it is more convenient for users of positional catalogues, the compilers usually express μ in terms of its components in RA and dec.; i.e., μ_α, μ_δ. If μ is not given, it can be derived from:

$$\mu = \sqrt{(15\,\mu_\alpha)^2 \cos^2\delta + \mu_\delta^2} \tag{5.6}$$

where μ_α is in seconds of time and μ_δ in seconds of arc.

Consider μ_α and μ_δ for Polaris at different epochs:

	μ_α	μ_δ
1900.0	$+0^s.144\,0$	$+0''.001$
1950.0	$+0^s.181\,1$	$-0''.004$
1978.0	$+0^s.208\,1$	$-0''.008$

Using these values in Eqn. 5.6, μ is found, in all three cases, to be $0''.046\,2$.

φ can be determined from:

$$\left.\begin{aligned} \mu_\delta &= \mu \cos\varphi \\ 15\mu_\alpha &= \mu \sec\delta \sin\varphi \end{aligned}\right\} \tag{5.7}$$

From the first equation of Eqn. 5.7, substituting the value of μ_δ for 1900.0, φ can be found from

$$\cos\varphi = \frac{\mu_\delta}{\mu} = \frac{0.001}{0.046\,2};$$

i.e., $\varphi_{1900} = 88^\circ.7597$.

Again, at 1950.0, $\cos\varphi = \dfrac{-0.004}{0.046\,2}$ and $\varphi_{1950} = 94^\circ.966\,9$. For 1978.0, φ is found to be $99^\circ.971\,6$. Notice that φ is increasing; when it was exactly 90° the proper motion

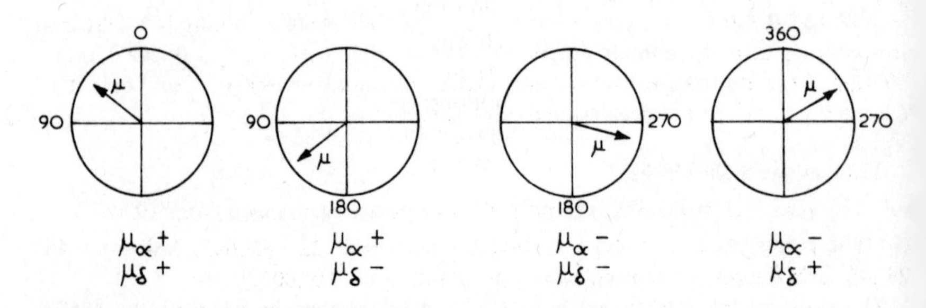

Fig 1 Correct quadrant for φ

in dec. would have been zero, and when it passed the value of 90° the sign of μ_δ changed from positive to negative.

Care must be taken to establish the correct quadrant when employing Eqn. 5.7 if an electronic calculator is used. When $0^\circ < \varphi \leqslant 180^\circ$ the calculator will give the correct value for φ if the first equation of Eqn. 5.7 is used, because in the first quadrant the sign of $\cos\varphi$ is positive, in the second quadrant negative. But the sign of $\sin\varphi$ is positive in both cases, and the calculator will display only the first quadrant value if the second equation of Eqn. 5.7 is employed. To illustrate this, take any angle between 90° and 180°—say, 150°. Key this angle into the calculator and take the sine (0.500 0); now, if the $\sin^{-1}$ key is depressed, the display will show 30°. Now try with the cosine; key in 150°, take the cosine (–0.866 0) and press $\cos^{-1}$. The display shows the correct angle, 150°. But when $\varphi > 180^\circ$ the calculator will give incorrect solutions in both cases.

To overcome this difficulty, consult Fig. 1, which shows the correct quadrant of φ according to the signs of μ_a and μ_δ.

Users of HP calculators with facilities for polar to rectangular conversion can overcome this problem of quadrant in a simple manner. From Eqn. 5.7 we can deduce:

$$\tan\varphi = \frac{15\,\mu_a \cos\delta}{\mu_\delta} \tag{5.8}$$

Evaluate $15\,\mu_a \cos\delta$, then enter μ_δ. The numerator is now in the Y register of the stack and the denominator is in the X register. Now, instead of pressing ÷, key g → P, R ↓. This will give the angle φ, and its value will lie between –180° and +180°. If the display is negative, add 360°.

For further practice, try the following:

(c) Given $\mu_a = -0^s.024\,0$, $\mu_\delta = +0''.003$ for a star of $\delta = +32^\circ\,15'$, find μ and φ.

(d) Given, for a star where $\delta = +42^\circ\,30'$, $\mu = 0''.287\,2$, $\varphi = 47^\circ.356\,6$, find μ_a, μ_δ.

Your answers should be:

(c) $\mu = 0''.304\,5$, $\varphi = 270^\circ.564\,5$.

(d) $\mu_a = +0^s.019\,1$, $\mu_\delta = +0''.195$.

NOTES

6 Sun, Moon and Planets

The reader of this chapter must face up to two inescapable facts.

The first is that, although the *mean* period of any planet or satellite can be established from a long series of observations, this information does not enable the precise geocentric coordinates of any of the bodies to be found for any particular time. This would only be possible if the Sun were accompanied by a single planet, which in turn did not have any satellites. The reason is, of course, that the planets interact gravitationally with one another as well as with the Sun as they move in their individual orbits. The masses of the planets are unequal, their orbits are not circular and, being at different distances from the Sun, they have different periods. It follows that the perturbative forces mutually exerted vary greatly with the times and, owing to the several factors involved, are very difficult to predict accurately. Difficult, but not impossible.

The second unpalatable fact is that, although the accelerative effects of the perturbations are not impossible to predict, it must be admitted that the task is beyond the normal computational resources afforded by the types of calculator with which this book is concerned, and the stated aim of accuracy with speed.

Why, then, include a chapter with this heading? The answer is simply that I do not like glossing over unwelcome facts, and the reader is entitled to know why such and such a type of computation is so time-consuming and intricate that it is best left well alone for the experts to perform and tabulate in the *AE*.

However, there is a ray of hope. USNO Circular No. 155 (October 1976) is an *Almanac for Computers for the year 1977*. By means of tables of Chebyshev coefficients it enables close approximations of the coordinates to be made quickly, with a choice of level of accuracy, for *any* time during the year (e.g., the RA and dec. of the Sun at $13^{h}\ 31^{m}\ 00^{s}$ UT on 1977, May 7).* Perhaps the compilers could be persuaded to publish, in one volume, similar tables for Sun, Moon and planets only, for a 10-year period, or even up to epoch 2000.0. There is some speculation that with the rising cost of publication the *AE* as we know it may become prohibitively expensive, and it may well be that astronomers might prefer to compute for themselves the ephemerides in which they have a particular interest.

* A similar publication for 1978 has been issued in Paris by the *Bureau des Longitudes* under the title *Connaissance des Temps, Nouvelle Série, Ephémérides pour l'An* 1978.

The technique employed would undoubtedly make it much easier for observers to compute, quickly and accurately, the equatorial coordinates for any desired time and thus avoid the need for Besselian interpolation between tabulated ephemerides. But until such a publication becomes more generally available to computers the route to accuracy will remain through Newcomb's tables (*Astronomical Papers of the American Ephemeris*, Vol. VI, 1898) and the *Improved Lunar Ephemeris*. There are drawbacks, too. Chebyshev coefficients by themselves tell you nothing; but by looking at the tabulated coordinates in the *AE* one can see at a glance when interesting events (such as conjunctions) are going to occur. All this having been said, it is still possible to compute reasonably accurate coordinates for the Sun and most of the planets throughout the year. The approximate methods are demonstrated in Chapter 9.

Programmes for use with Chebyshev coefficients are included in Appendix II (Programmes 51 to 53).

7 Visual Binary Star Orbits

Topic 1 Elements of the orbit of a visual binary star.
Topic 2 Position angle and separation at any epoch.

1 **To compute the elements of the orbit of a visual binary star, P, T, e, a, ω, i, Ω, where**

$P =$ the period of revolution in mean solar years
$T =$ the time of periastron passage
$e =$ the numerical eccentricity of the orbit
$a =$ the major semi-axis, expressed in seconds of arc
$\omega =$ the angle in the plane of the true orbit between the line of nodes and the major axis, measured from the nodal point Ω to the point of periastron passage in the direction of the companion's motion (the value can be anywhere between 0° and 360°)
$\Omega =$ the position of the nodal point which lies between 0° and 180°, assumed to be the ascending node (see note below) (the other nodal point does not enter into the computation, so when *the* nodal point is referred to it means Ω)
$i =$ the inclination of the orbit plane; the value lies between 0° and 180° (see note below); direct motion of the companion (position angles increasing) is indicated by $0° < i \leqslant 90°$, retrograde motion (position angles decreasing) by $90° < i \leqslant 180°$.

Note: Measurements of the position angle and separation provide information only about the *apparent* orbit, which lies in the plane perpendicular to the line of sight. In these circumstances it is not possible to establish which of the nodes is actually the ascending node. It is conventional to select a value for Ω less than 180°, unless radial-velocity measurements of the companion give an indication of the true inclination of the orbit. The computed value for i is often shown as $\pm$ until the indeterminacy of i and Ω is removed by such radial-velocity measures. When these are available, i is taken to be positive if the orbital motion at the nodal point is taking the companion away from the observer, or negative if the motion is toward the observer at this point in the orbit.

Ω is measured with respect to the pole at a specified epoch; it follows that, owing to precession, Ω (and consequently ω) will change slowly with time. This aspect is covered in Topic 2.

The equations: There are several interdependent equations involved in the computation of an orbit. In this topic I have thought it more appropriate to introduce these as required in the working of the example.

Further information: R. G. Aitken, *The Binary Stars*, Chap. 4; D. McNally, *Positional Astronomy*, Chap. 12.3.

Method of calculation: The Thiele-Innes method is illustrated here. The working has been broken down into several logical steps; in each step, any equations to be used are given, followed by the working. It will be found that the calculation at each step is relatively short and straightforward; therefore, no distinction between algebraic and RPN calculators has been made as it is unnecessary for the key-strokes to be listed.

Example: The following measures of a very close visual binary, 24 Aquarii (*ADS* 15 176), are given. They were made in the interval 1890–1932, and have been taken from the table on p 103 (Dover paperback edition) of Aitken's *The Binary Stars*. Only those measures made with telescope apertures of 24 inches and over have been selected as an illustrative example. There are, of course, many later measures (see the 'further practice' problems at the end of the chapter) which we would use if we were attempting to compute a definitive orbit, but this will not be our objective; the aim must be restricted to that of showing how it is done.

24 Aquarii provides a good example of a difficult case: the orbit is highly eccentric, the pair always much less than a second of arc apart.

Date	p.a.	d	n
	°	″	
1890.75	254.5	0.45	3
1.75	261.0	0.55	4
2.40	256.2	0.38	2
3.88	262.8	0.59	1
4.82	264.7	0.52	7
7.81	263.5	0.65	3
7.89	267.4	0.73	1
8.78	269.0	0.49	3 (incl. 12″)
8.84	269.0	0.54	1
1901.54	269.4	0.49	10
1.79	274.0	0.55	2
4.67	278.6	0.49	1
8.72	279.6	0.68	2
8.72	284.8	0.56	2
8.73	286.4	0.72	2
1910.72	278.2	0.43	5
4.00	292.5	0.47	8
4.63	291.3	0.47	2
4.66*	293.5	0.51	1
6.42	296.5	0.53	3
7.74	294.7	0.42	1
1921.66	321.1	0.22	3

4.55	55.0	0.12	1
4.71	6.9	0.22	1
4.82	350.0	0.16	1
6.64	190.7	0.20	1
6.69	204.2	0.19	1
7.74	211.0	0.21	3
7.74	218.7	0.23	1
8.73*	224.6	0.26	1
8.75	221.2	0.27	4
8.75	222.6	0.26	4
9.46*	228.8	0.28	4
9.63*	230.2	0.27	1
9.86*	227.8	0.26	3
1930.48*	234.3	0.29	3
1.66*	236.0	0.37	2
2.79*	236.9	0.35	4
2.79*	238.3	0.30	1

By definition, the position angle of a double star is measured in degrees, with 0° indicating the North point; that is, the direction of the North Celestial Pole. But, as we have seen earlier, the NCP is subject to slow displacement due to precession. It therefore follows that, over an extended period of time, position angle measurements will relate to different North points.

When assembling material it is usual to consider the effect of precession on measures of position angle spread over many years. Where the star is not near the pole and its annual proper motion is small, then the effect of precession will be negligible unless the period over which the measures are spread is lengthy.

As a guide, the correction applied to a position angle for a star of declination between 30° and 40°, with little or no proper motion in RA, will be approximately 0°.2 over a period of some 20 years. No correction is necessary for the separation measures. See the introduction to Topic 2 of this Chapter for details of the correction to be applied. In this worked example we shall proceed on the assumption that any such corrections have already been carried out, and that the position angles are referred to a standard epoch, 1900.0, although in practice, because of its position on the celestial equator, no correction for 24 Aquarii would be required.

1. First, we plot these measures on graph paper, with time along the x-axis, from 1890 to 1935, position angles and distances to convenient scales along the y-axis. Position angles are marked with dots, graduated in size relative to the number of nights observed, n; distances are marked with crosses, again relative in size to n. The size of the plot marks therefore gives a direct indication of the weighting to be favoured where measures are discordant.

Note that, in order to save space, the position-angle plot has been split into two segments, the one on the left for the period 1890 to 1921.66, the one on the right from 1926.64 to 1933; the section near periastron where the measured angles are difficult and therefore discordant has been omitted.

* Asterisks indicate measures not available to W. S. Finsen in the calculation of the orbit as published in 1929, and therefore in the example used in Aitken's book.

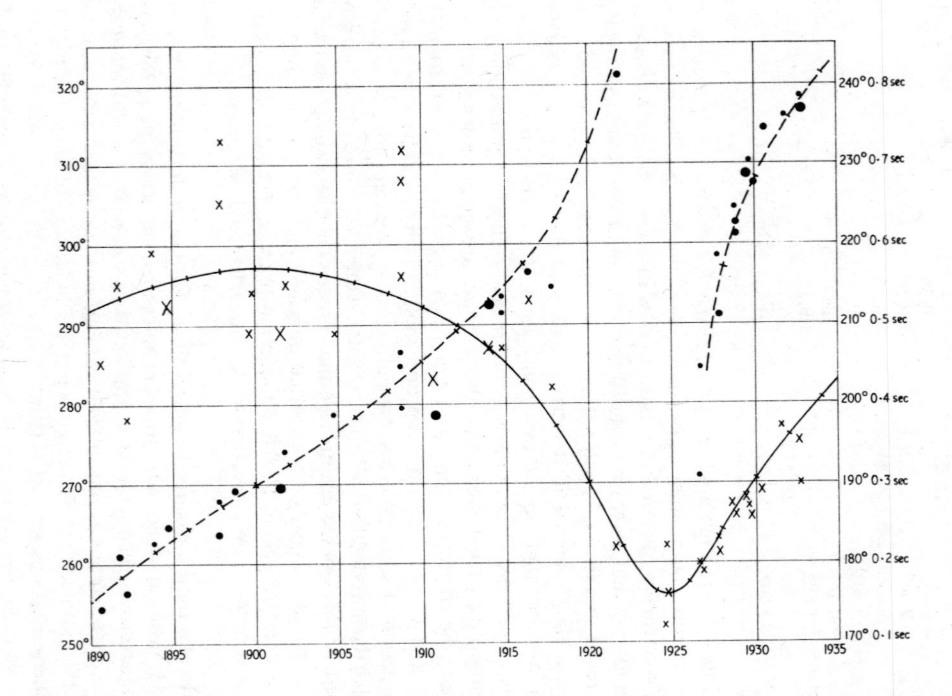

Fig 2. Graphical representation of position angle and separation of 24 Aqr (ADS 15176) during the period 1890–1935

Smooth arcs are then drawn through the plotted points, favouring the larger marks where possible, and taking care to ensure that the completed graph properly reflects Kepler's second law, in particular that the rate of change in position angle is greatest at periastron, and in synchronism with the dip in the distance curve. Distance is shown with a solid line, position angle by the two broken lines.

Where measures of binary pairs wider than than 2″ have been derived from multi-exposure photographic plates made with long-focus instruments they should be accorded greater weight than visual measures. (See P. van de Kamp, *Principles of Astrometry*, Chap. 10.2, 10.3 and p 149.)

Note that the measures of position angle (dots) are reasonably consistent (except, of course, at periastron, where they have been omitted) enabling a fairly reliable smooth arc to be drawn through the plot, while the distance measures (crosses) are often discordant. In principle, this shows that visual measures of position angle are easier to make than measures of the separation of the two stars, and especially so in the case of very close binaries such as 24 Aquarii. But notice, also, at the very time when one would expect the distance measures to become easier, at apastron, in this example they fluctuate widely, from about 0″.45 to 0″.7, in spite of the fact that only those measures made with large-aperture telescopes have been used.

At periastron (by inspection seen to occur at about 1924.75), and thereafter up to 1933, the measures of distance become much more reliable. There seems to be no reasonable explanation why this improvement, after about 1915, should be so dramatic; the possibility that the later measures, made in the knowledge of what other observers were recording, and therefore the known trend of the orbital motion, were coloured by what observers expected to see rather than what was actually seen can, I think, fairly be discounted on the grounds of the professional stature of these observers whose measures have been used, and the size and efficiency of the telescopes and micrometers they employed. One hopes, too, that such distortion will not occur in any computation based on those observations.

2. Next, each curve is marked with cross-ticks at 2-yearly intervals from 1890, so that average measures can be taken at even increments of time for tabulation. When the table is completed, the rates of change in p and d are compared with the graph. The cross-ticks should show immediately that at apastron, when they are grouped closer together, p and d change only slowly; at periastron the wider gaps between adjacent ticks reflect the acceleration and subsequent deceleration of the companion as it swings around the primary at closest approach. Note, too, that the slope of p begins to increase synchronously with the marked dip in d; also, the downward change of slope at the extreme left of the graph for p is in agreement with the line for d. We are thus reassured that no fundamental laws of astro-dynamics have inadvertently been transgressed, and that the prepared curves are as accurate a graphical representation of the orbital changes as can be deduced from the published measures.

1 Year	2 p	3 Δp	4 d	5 $d_1 d_2 \Delta p$	6 p	7 Δp	8 d	9 $d_1 d_2 \Delta p$
	°	°	″		°	°	″	
1890	255.5		0.52		255.5		0.53	
		+3.1		+0.85		+3.2		+0.92
92	258.6		0.53		258.7		0.54	
		2.9		0.85		3.0		0.89
94	261.5		0.55		261.7		0.55	
		3.0		0.92		2.9		0.89
96	264.5		0.56		264.6		0.56	
		3.0		0.96		2.8		0.89
98	267.5		0.57		267.4		0.57	
		2.2		0.71		2.7		0.88
1900	269.7		0.57		270.1		0.57	
		2.8		0.91		2.8		0.91
02	272.5		0.57		272.9		0.57	
		2.8		0.89		2.9		0.93
04	275.3		0.56		275.8		0.56	
		3.1		0.95		3.0		0.92
06	278.4		0.55		278.8		0.55	
		3.5		1.04		3.2		0.93
08	281.9		0.54		282.0		0.53	
		3.5		0.98		3.4		0.92
1910	285.4		0.52		285.4		0.51	
		3.7		0.96		3.6		0.90
12	289.1		0.50		289.0		0.49	
		4.3		1.01		4.0		0.90
14	293.4		0.47		293.0		0.46	
		4.3		0.87		4.6		0.89
16	297.7		0.43		297.6		0.42	
		5.8		0.92		5.7		0.89
18	303.5		0.37		303.3		0.37	
1928	217.3		0.24		216.2		0.24	
		10.9		0.78		12.3		0.89
1930	228.2		0.30		228.5		0.30	
		7.5		0.83		8.0		0.89
32	235.7		0.37		236.5		0.37	
				Mean 0.9019				Mean 0.9025

$$c = \frac{\text{mean}}{2 \times 57.295\,78} \tag{7.1}$$

where c is the double areal constant, 2 is the interval of time in years between successive values, and 57.295 78 converts Δp into radians.

$c = \dfrac{0.9025}{114.591\,56} = +0.007\,876$. W. S. Finsen found $+0.007\,81$; Aitken found $+0.007\,914$.

Inspection of the columns of the table produced from these curves should confirm that the motions they represent are in agreement with the curves and orbital theory. Note that in this method the need for a complete graphical construction of the ellipse is avoided.

Columns 1, 2 and 4 to the left of the double line are completed from the graph. Column 3 is then completed to show the 2-yearly differences in Column 2. Column 5 is completed by multiplying Δp by the values of d immediately above and below it; the first entry is thus $+3.1 \times 0.52 \times 0.53 = +0.85$.

Examination of this left-hand section reveals the need for adjustment in order to smooth out the rates of change. Over the short period of two years one may safely consider the $d_1 d_2 \Delta p$ figures in Column 5 as a fair representation of the double areas of the sectors swept out, and they should therefore be reasonably constant. In fact, the values in Column 5 span the range 0.71 to 1.04, and can be improved by adjusting either or both p and d in Column 6 and 8. In general, it will be found necessary to effect the bulk of the adjustments in d, and this will be evident from the graph, where it is obvious that the position angle measures are more reliable.

Great care is required in making these adjustments; in deciding where they are desirable, and in the amount of adjustment applied. The end result to aim for is:
(*i*) close agreement of all the figures in Column 9, the mean of which should be close to the mean of Column 5;
(*ii*) the top and bottom values in Columns 6 and 8 to be in close agreement with those in 2 and 4 respectively;
(*iii*) the rates of change in Column 7 to be smooth and to agree with the expected rate of acceleration or deceleration of the companion in its orbital path;
(*iv*) the values in Column 8 to reflect a smooth curve as the distance from the primary changes due to orbital motion.

The initial process has been explained at length with the purpose of achieving the greatest objectivity in carrying out any adjustments to the observer's own records and other published observations. The rest of the computation of the orbit may then confidently be based on the firm foundation of observational material which is in fair accord with gravitational theory, and yet which has not been doctored so ruthlessly to fit the expected motion as to be no more than a figment of the imagination.

3. From the table we select three 'normal' places, spaced as widely apart on the observed portion of the ellipse as is convenient, and avoiding any potential trouble-spots such as the part of the orbit near periastron, where measures are likely to be more discordant than elsewhere. Accordingly, we choose:

(1)	1890	255°.5	0″.53
(2)	1910	285°.4	0″.51
(3)	1930	228°.5	0″.30

These normal places are different from those used in Aitken's example of the Thiele-Innes method of computation. It will be instructive to compare the eventual result with that of Finsen and also with that of Danjon in the *Catalogue of Visual Binary Orbits* (Publications of the USNO, Vol. XVIII, Part III), based on later measures.

4. From $x = d\cos p$, (7.2)

$y = d\sin p$, (7.3)

$\Delta_{1,2} = x_1y_2 - x_2y_1$, etc. (7.4)

we find:

$t_1 = 1890$	$x_1 = -0.132\,7$	$y_1 = -0.513\,1$	$\Delta_{1,2} = +0.134\,7$,
$t_2 = 1910$	$x_2 = +0.135\,4$	$y_2 = -0.491\,7$	$\Delta_{2,3} = -0.128\,2$,
$t_3 = 1930$	$x_3 = -0.198\,8$	$y_3 = -0.224\,7$	$\Delta_{1,3} = -0.072\,2$.

(*Note:* with HP calculators, the x, y coordinates may be found quickly by: p, ↑ , d, $f \rightarrow$ R. x and y are then in the X and Y registers. x is displayed; to see y, key $x \leftarrow\rightarrow y$).

Then, using the double areal constant c from Eqn. 7.1 (in this case, +0.007 876):

$$\frac{\Delta_{1,2}}{c} = +17.10$$

$$\frac{\Delta_{2,3}}{c} = -16.28$$

$$\frac{\Delta_{1,3}}{c} = -9.17$$

From $$t_2 - t_1 - \frac{\Delta_{1,2}}{c} = \frac{1}{\mu}(u - \sin u) \qquad (7.5)$$

and substituting

$1910 - 1890 - 17.10 = 2.90$

we find $u - \sin u = 2.90\mu$;

from $$t_3 - t_2 - \frac{\Delta_{2,3}}{c} = \frac{1}{\mu}(v - \sin v) \qquad (7.6)$$

and substituting

$1930 - 1910 + 16.28 = 36.28$

we find $v - \sin v = 36.28\mu$;

and from $$t_3 - t_1 - \frac{\Delta_{1,3}}{c} = \frac{1}{\mu}[(u + v) - \sin(u + v)] \qquad (7.7)$$

and substituting

$1930 - 1890 + 9.17 = 49.17$

we find $(u + v) - \sin(u + v) = 49.17\mu$.

5. We now have to construct a table of approximations for μ. With the aid of our electronic calculators this task will be greatly speeded up. Inspection of the graph will often enable a rough estimate of the period of the orbit to be made, and this will be facilitated if both ends of the ellipse can be identified. Deduct the time of approximate apastron from the following time of approximate periastron, and multiply by 2, as an estimate of μ.

From: $$P = \frac{2\pi}{\mu} \qquad (7.8)$$

we see $$\mu = \frac{2\pi}{P} \qquad (7.9)$$

$$= \frac{6.283\,185\,3}{(1924.75 - 1900) \times 2}$$
$$= 0.127.$$

Thus, we can expect μ to lie somewhere in the range 0.12 to 0.13.

1.	2.	3.
$\mu =$	0.12	0.13
(*i*) $u - \sin u = 2.90\mu$	0.348 0	0.377 0
(*ii*) $v - \sin v = 36.28\mu$	4.353 6	4.716 4
(*iii*) $(u+v) - \sin(u+v) = 49.17\mu$	5.900 4	6.392 1
(*iv*) u	1.315 6 *r* =75°.378 3	1.353 5 *r* =77°.549 8
(*v*) v	3.767 65 *r* =215°.870 4	3.975 7 *r* =227°.780 7
(*vi*) Sum (degrees)	291°.248 7	305°.330 6
(*vii*) $(u+v)$	4.922 4 *r* =282°.032 7	7.162 2 *r* =410°.363 8
(*viii*) Difference (degrees)	+9°.216 0	–105°.033 2

4.	5.	6.	7.
0.122	0.124	0.123	0.123 3
0.353 8	0.359 6	0.356 7	0.357 6
4.426 2	4.498 7	4.462 4	4.473 3
5.998 7	6.097 1	6.047 9	6.062 7
1.323 4 *r* =75°.825 2	1.331 0 *r* =76°.260 7	1.327 2 *r* =76°.043 0	1.328 4 *r* =76°.111 7
3.808 0 *r* =218°.182 3	3.848 9 *r* =220°.525 7	3.828 35 *r* =219°.348 3	3.834 5 *r* =219°.700 7
294°.007 5	296°.786 4	295°.391 3	295°.812 4
5.057 7 *r* =289°.784 9	5.226 2 *r* =299°.439 2	5.136 5 *r* =294°.299 8	5.162 2 *r* =295°.772 3
+4°.222 6	–2°.652 8	+1°.091 5	+0°.040 1

Start in column 2 of the table by taking the trial value of 0.12 for μ. Lines (*i*), (*ii*) and (*iii*) are completed thus:

(*i*) $0.12 \times 2.9 = 0.348\,0$
(*ii*) $0.12 \times 36.28 = 4.353\,6$
(*iii*) $0.12 \times 49.17 = 5.900\,4$

No more than four places of decimals are normally required. Line (*iv*) is computed, in the absence of a table for $x - \sin x$, by iteration, evaluating u first in radians, then converting into degrees, thus:

> Switch calculator to function in radian mode. Make trial $u = 1$ rad. Enter 1 in the memory, then f sin; exchange x-display and memory; deduct the figure stored in the memory from that in the X-register.

Those with a programmable calculator can key in this set of instructions:*

Switch to Programme
g RAD

* As written, suitable for the HP-25. Slight modification will be necessary for other models.

```
STO 0
f sin
f last x
x <--> y
-
GTO 00
Switch to Run
f PRGM
enter 57.295 78 in STO 1
enter trial x
R/S
```

Read result and re-iterate, until the display agrees with line (*i*). Note this result, then RCL 1, ×, to convert into degrees.

This programme will suffice for short computations, but those who would prefer to employ a more sophisticated method which automatically reiterates until the desired result is found, and then displays it, may like to try the iteration programme for $x - \sin x$ in the Appendix. With slight amendments, this would be specially suited to calculators with magnetic-card facilities. (See Programme 14.)

Now, back to the manual computation. The first trial for $u = 1$ rad gives $u - \sin u = 0.158\,5$. This is too low (in line (*i*), $u - \sin u = 0.348\,0$). Try $u = 2$ rad. The result is 1.090 7. Too high! Reiterate between 1 and 2 rads until the result agrees with line (*i*). The value for u is eventually found to be 1.315 6 rad $= 75^\circ.378\,3$. Enter these values in Column 2 against line (*iv*). Line (*v*) is the result of a similar iterative process, this time for v. The value for v which gives $v - \sin v$ in agreement with line (*ii*) is 3.767 65 rad $= 215^\circ.870\,4$. Line (*vi*) is line (*iv*) + line (*v*), in degrees. Line (*vii*) is yet another iteration, this time of line (*vi*), expressed in radians, to a value which agrees with line (*iii*): i.e., 5.900 4. In line (*vi*) we found, for the sum of u and v from lines (*iv*) and (*v*), $291^\circ.248\,7 = 5.083\,2$ rad. Take this as a trial value for $(u + v)$ and iterate for $(u + v) - \sin(u + v)$ in exactly the same manner as the previous iterations. The trial value gives 6.015 3. This is too high, as line (*iii*) has been evaluated at 5.900 4. Try 4.9, which gives 5.882 5—too low. Reiterate between 4.9 and 5.0 rad, until agreement with line (*iii*) is reached. The value for $(u + v)$ at which agreement is achieved is 4.922 4 rad. Convert into degrees, $282^\circ.032\,7$, and complete the entry for line (*vii*) in Column 2. Line (*viii*) is simply line (*vi*) minus line (*vii*), in degrees. We use line (*viii*) as a guide to the accuracy of the estimated value for μ at the head of the column, and aim to reduce this difference to zero, or nearly so.

So, we repeat the whole process in Column 3, this time taking $\mu = 0.13$. The final difference in line (*viii*) Column 3 is found to be $-105^\circ.033\,2$. Thus we have effectively bracketed the first two trial shots on either side of the true value, but evidently not evenly. Therefore, in Column 4, after considering the two line—(*viii*) entries, we set $\mu = 0.122$ and find that the difference in line (*viii*) has been reduced to $+4^\circ.222\,6$.

Successive trials in adjoining columns gradually bring the calculation of μ closer to its true value until in Column 7 the difference has been reduced to $+0^\circ.040\,1$

There is no point in proceeding further, and so we establish the value of μ as 0.123 3. Aitken found 0.122 4, and Finsen 0.122 41.

From Column 7 we extract:

$$\mu = 0.123\,3 \qquad u = 76^\circ.111\,7 \qquad v = 219^\circ.700\,7$$

(Those with programmable calculators will have found the whole iterative process quite easy to carry out.)

6. The period of the orbit, P, is established from Eqn. 7.8:

$$P = \frac{2\pi}{\mu} = \frac{6.283\,19}{0.123\,3} = 50.96 \text{ years;}$$

and the mean annual motion, n, from

$$n = 57.295\,78\,\mu = 7^\circ.064\,6.$$

7. The eccentric anomaly E for each of the three normal places, and e, the numerical eccentricity of the orbit, are computed as follows.

From
$$e\sin E_2 = \frac{(\Delta_{2,3}\sin u) - (\Delta_{1,2}\sin v)}{\Delta_{1,2} + \Delta_{2,3} - \Delta_{1,3}} \tag{7.10}$$

and
$$e\cos E_2 = \frac{(\Delta_{2,3}\cos u) + (\Delta_{1,2}\cos v) - \Delta_{1,3}}{\Delta_{1,2} + \Delta_{2,3} - \Delta_{1,3}} \tag{7.11}$$

we obtain (*i*) $e\sin E_2 = -0.4880$ and
(*ii*) $e\cos E_2 = -0.7905$.

Dividing (*i*) by (*ii*) we have $\dfrac{\sin E_2}{\cos E_2} = +0.6174;$

that is, $\tan E_2 = +0.6174$,
and therefore $E_2 = 31^\circ.691\,9$ or $211^\circ.691\,9$.

As e is always positive, and both $e\sin E_2$ and $e\cos E_2$ are negative, it follows that E_2 must be in the third quadrant.

$$\therefore \quad E_2 = 211^\circ.691\,9, \text{ and } e = \frac{e\sin E_2}{\sin E_2} = \frac{-0.488\,0}{-0.525\,4} = 0.929\,0;$$

$$E_1 = E_2 - u = 211^\circ.691\,9 - 76^\circ.111\,7 = 135^\circ.580\,2;$$
$$E_3 = E_2 + v = 211^\circ.691\,9 + 219^\circ.700\,7 = 431^\circ.392\,6 - 360 = 71^\circ.392\,6.$$

(If your calculator features polar to rectangular conversion, see the note regarding the determination of quadrant at the end of Chapter 5.)

8. From Kepler's equation, the mean anomaly M can be derived from

$$M = E - e\sin E \tag{7.12}$$

where E is expressed in radians.

$$M_1 = 2.366\,3 - 0.929\,(+0.699\,9) = 1.716\,1 \text{ rad} = 98^\circ.325\,5;$$
$$M_2 = 3.694\,7 - 0.929\,(-0.525\,4) = 4.182\,8 \text{ rad} = 239^\circ.655\,2;$$
$$M_3 = 1.246\,0 - 0.929\,(+0.947\,7) = 0.365\,6 \text{ rad} = 20^\circ.947\,2.$$

9. T, the time of next periastron passage, is obtained from:

$$T = t + \left(P - \frac{M}{n}\right) \tag{7.13}$$

where t is the epoch of M, M is expressed in degrees, and P and n are as previously defined.

Then, $$T = 1890 + \left(50.96 - \frac{98.325\,5}{7.064\,6}\right) = 1927.04;$$

$$T = 1910 + \left(50.96 - \frac{239.655\,2}{7.064\,6}\right) = 1927.04;$$

$$T = 1930 + \left(50.96 - \frac{20.947\,2}{7.064\,6}\right) = 1977.995 - 50.96 = 1927.035.$$

We take T as 1927.04.

10. From $$X = \cos E - e \tag{7.14}$$
$$Y = \cos\varphi \sin E \tag{7.15}$$

where $\varphi = \sin^{-1}e$, and φ and E are expressed in degrees, find the X, Y pairs for each of the three normal places:

$$X_1 = -1.643\,2 \qquad Y_1 = +0.259\,0$$
$$X_2 = -1.779\,9 \qquad Y_2 = -0.194\,4$$
$$X_3 = -0.609\,9 \qquad Y_3 = +0.350\,7.$$

11. Evaluate the Innes constants A, B, F, G, from the first and third normal X, Y pairs, from

$$x = AX + FY \tag{7.16}$$
$$y = BX + GY \tag{7.17}$$

where x and y are as found in Step 4.

$$x_1 = -0.132\,7 = -1.643\,2\,A + 0.259\,0\,F \tag{a}$$
$$x_3 = -0.198\,8 = -0.609\,9\,A + 0.350\,7\,F \tag{b}$$

To eliminate F, multiply (b) by $\dfrac{-0.259\,0}{0.350\,7} = -0.738\,5.$

$$+0.146\,8 = +0.450\,4\,A - 0.259\,0\,F \tag{c}$$
Add (a) $$-0.132\,7 = -1.643\,2\,A + 0.259\,0\,F \tag{a}$$
$$+0.014\,1 = -1.192\,8\,A \tag{d}$$

$$A = \frac{0.014\,1}{-1.192\,8} = -0''.011\,8$$

Now solve for B.

$$y_1 = -0.513\,1 = -1.643\,2\,B + 0.259\,0\,G \tag{e}$$
$$y_3 = -0.224\,7 = -0.609\,9\,B + 0.350\,7\,G \tag{f}$$

To eliminate G, multiply (f) by the same factor as before, $-0.738\,5$.

$$+0.165\,9 = +0.450\,4\,B - 0.259\,0\,G \tag{g}$$
Add (e) $$-0.513\,1 = -1.643\,2\,B + 0.259\,0\,G \tag{e}$$
$$-0.347\,2 = -1.192\,8\,B \tag{h}$$

$$B = \frac{-0.347\,2}{-1.192\,8} = +0''.291\,1.$$

Solve for F:

Take (a) $x_1 = -0.132\,7 = -1.643\,2\,A + 0.259\,0\,F$ (a)

and (b) $x_3 = -0.198\,8 = -0.609\,9\,A + 0.350\,7\,F.$ (b)

This time, eliminate A. Multiply (b) by $-\dfrac{-1.643\,3}{-0.610\,0} = -2.694\,2.$

$$+0.535\,6 = +1.643\,2\,A - 0.944\,9\,F \quad (i)$$

Add (a) $$-0.132\,7 = -1.643\,2\,A + 0.259\,0\,F \quad (a)$$

$$+0.402\,9 = -0.685\,9\,F \quad (j)$$

$$F = \frac{0.402\,9}{-0.685\,9} = -0''.587\,4$$

Lastly, solve for G:

Take (e) $y_1 = -0.513\,1 = -1.643\,2\,B + 0.259\,0\,G$ (e)

and (f) $y_3 = -0.224\,7 = -0.609\,9\,B + 0.350\,7\,G$ (f)

This time, eliminate B. Multiply (f) by the same factor as in the solution for F, $-2.694\,2$.

$$+0.605\,4 = +1.643\,2\,B - 0.944\,9\,G \quad (k)$$

Add (e) $$-0.513\,1 = -1.643\,2\,B + 0.259\,0\,G \quad (e)$$

$$+0.092\,3 = -0.685\,9\,G \quad (l)$$

$$G = \frac{0.092\,3}{-0.685\,9} = -0''.134\,6.$$

12. Check the values found for A, B, F, G, against the second normal place:

$$x_2 = AX_2 + FY_2$$
$$y_2 = BX_2 + GY_2$$
$$x_2 = (-0.011\,8)(-1.779\,9) + (-0.587\,4)(-0.194\,4) = +0.135\,2$$
$$y_2 = (+0.291\,1)(-1.779\,9) + (-0.134\,6)(-0.194\,4) = -0.492\,0$$

From Step 4, $x_2 = +0.135\,4$, $y_2 = -0.491\,7$, a satisfactory check. The slight discrepancy could have been avoided by working to an extra place of decimals in the preceding calculations but this extra work is not really justified.

13. Tabulate:

$A =$	$-0.011\,8$	$B =$	$+0.291\,1$
$G =$	$-0.134\,6$	$F =$	$-0.587\,4$
$A + G =$	$-0.146\,4$	$B - F =$	$+0.878\,5$
$A - G =$	$+0.122\,8$	$-B - F =$	$+0.296\,3$

Evaluate ω, Ω, and i, defined at the beginning of the topic, from

$$\tan(\omega + \Omega) = \frac{B - F}{A + G} \qquad (7.18)$$

$$\tan(\omega - \Omega) = \frac{-B - F}{A - G} \qquad (7.19)$$

$$\tan^2\frac{i}{2} = \frac{-B - F}{B - F} \times \frac{\sin(\omega + \Omega)}{\sin(\omega - \Omega)}. \qquad (7.20)$$

Before solving for ω and Ω we must first resolve the matter of quadrants. It is not difficult to establish in which quadrants $(\omega + \Omega)$ and $(\omega - \Omega)$ lie. Consider the Thiele-Innes equation:

$$A + G = 2\,a\cos(\omega + \Omega)\cos^2\frac{i}{2} \tag{7.21}$$

We can say: $A + G$ is negative; a is always positive; and it does not matter if $\cos\frac{i}{2}$ is negative or positive because, either way, its square will be positive.

Therefore, to make $A + G$ negative, $\cos(\omega + \Omega)$ must also be negative. (Remember: 1st quadrant all positive, 2nd quadrant only sine positive, 3rd quadrant only tangent positive, 4th quadrant only cosine positive.)

Now consider:

$$B - F = 2\,a\sin(\omega + \Omega)\cos^2\frac{i}{2} \tag{7.22}$$

Following the same line of reasoning, we can conclude that, because $B - F$ is positive, $\sin(\omega + \Omega)$ must also be positive.

From $$A - G = 2\,a\cos(\omega - \Omega)\sin^2\frac{i}{2} \tag{7.23}$$

we can say $\cos(\omega - \Omega)$ is positive.

From $$-B - F = 2\,a\sin(\omega - \Omega)\sin^2\frac{i}{2} \tag{7.24}$$

we can say $\sin(\omega - \Omega)$ is positive.

Assembling these conclusions:

$\cos(\omega + \Omega)$ is negative; $\sin(\omega + \Omega)$ is positive. It follows that $(\omega + \Omega)$ lies in the second quadrant.

$\cos(\omega - \Omega)$ is positive; $\sin(\omega - \Omega)$ is positive. It follows that $(\omega - \Omega)$ lies in the first quadrant.

Now proceed with Eqns. 7.18 and 7.19.

$$\tan(\omega + \Omega) = \frac{+0.878\,5}{-0.146\,4} = -6.000\,7$$

$$\tan(\omega - \Omega) = \frac{+0.296\,3}{+0.122\,8} = +2.412\,9$$

$\omega + \Omega = 99^\circ.461\,3$ (second quadrant)

$\omega - \Omega = 67^\circ.488\,7$ (first quadrant)

Add for $2\omega = 166^\circ.950\,0$

Subtract for $2\Omega = 31^\circ.972\,6$

Thus $\omega = 83^\circ.475$; $\Omega = 15^\circ.986$ (at 1900.0)

$\cos(\omega + \Omega) = -0.164\,4$; $\sin(\omega + \Omega) = +0.986\,4$

$\cos(\omega - \Omega) = +0.382\,9$; $\sin(\omega - \Omega) = +0.923\,8$

Then, in Eqn. 7.20:

$$\tan^2\frac{i}{2} = \frac{+0.296\,3}{+0.878\,5} \times \frac{+0.986\,4}{+0.923\,8} = +0.360\,1$$

$\therefore\ \tan\frac{i}{2} = \sqrt{0.360\,1} = \pm 0.6001\,1$. So $\frac{i}{2} = \pm 30^\circ.968\,5$, and $i = \pm 61^\circ.94$ (Position angles increasing with time—see note in the definition for i).

Note: if Ω were to be found in the fourth quadrant, then 180° would have to be added to (or subtracted from) *both* ω and Ω.

14. Finally, a, the major semi-axis, is found from

$$a = \frac{B - F}{2 \sin(\omega + \Omega) \cos^2 \frac{i}{2}} \qquad (7.25)$$

$$= \frac{+0.878\,5}{2 \times 0.986\,4 \times 0.735\,2}$$

$= 0''.605\,7$

$= 0''.61.$

Alternatively,

$$a = \frac{(A + G) \sec(\omega + \Omega)}{1 + \cos i} \qquad (7.26)$$

When using pocket calculators, secants are obtained by taking the reciprocal of the cosine, e.g., f cos, g $\frac{1}{x}$.

$$= \frac{(-0.146\,4) \times (-6.083\,4)}{1 + 0.470\,4}$$

$$= \frac{+0.890\,6}{+1.470\,4}$$

$= 0''.605\,7$

$= 0''.61$, agreeing with the value derived above.

15. Apply two checks, using the values found for ω, Ω, i and a, to prove back to the Thiele-Innes constants.

(i) $A = a(\cos\omega \cos\Omega - \sin\omega \sin\Omega \cos i)$ (7.27)

$= 0.605\,7\,[(0.113\,6 \times 0.961\,3) - (0.993\,5 \times 0.275\,4 \times 0.470\,4)]$

$= 0.605\,7\,(0.109\,2 - 0.128\,7)$

$= -0.011\,8$, which agrees with the value for A tabulated at the beginning of Step 13.

(ii) Use Eqn. 7.24 for the second check:

$-B - F = 2 \times 0.605\,7 \times 0.923\,8 \times 0.264\,8$

$= +0.296\,3$, which also agrees with the value tabulated in Step 13.

So the mathematical checks are satisfactory.

16. Assembling the elements of the orbit thus computed, we can then compare them with those previously published. Do not show more than one or two decimal places, as this would imply a degree of accuracy which is not justified.

	Computed		Finsen	Danjon
$P =$	50.96		51.33	48.7
$T =$	1927.04		1925.68	1923.01
$e =$	0.93		0.910 2	0.86
$a =$	0″.61		0″.525	0″.42
$\Omega =$	16°.0	1900.0	4°.95	139°.8
$\omega =$	83°.5		87°.35	295°.0
$i =$	±61°.9		±56°.02	55°.2

The epoch shown against our computed elements is a reminder only; it will be essential to do this for non-equatorial stars. This concludes the calculation of the orbit.

It should now be obvious that in the case of a very close and difficult binary such as 24 Aquarii the slightest variation in the interpretation of discordant measures, or in the weighting applied in respect of the proven reliability of the various observers, or in any adjustments made to eliminate systematic or personal errors, will have a great effect on the elements of the orbit subsequently computed, and particularly in the values found for ω, Ω and i.

It must also be stressed that this present calculation has been conducted only as an exercise in visual-binary-orbit computing techniques, and must *not* be interpreted or used as a definitive orbit. Remember, we chose to ignore the measures of some dedicated observers simply because the apertures of their telescopes were less than 24 inches; this is not a convincing scientific reason for discarding valuable research data.

To keep a sense of proportion about the subject, I can do no better than to refer the reader to W. D. Heintz in *Astronomy—A Handbook*, Ed. G. Roth, 20.4. He rightly points out that wherever sufficient orbital motion of a binary pair has been established, one or more orbits have already been calculated, and takes the view that new computations should not be encouraged. He criticizes "computer-happy people" for publishing redundant duplications which fail to improve upon previous orbits. Strong stuff, but justifiable in many cases. It is all a question of degree, of course, and in the case of a binary with relatively short period, once a complete revolution has been observed and measured fresh computation should be able to improve on a preliminary orbit calculated from an arc.

There is clearly a moral obligation on the part of the computer unequivocally to ensure that (a) the bulk of the information upon which the orbit computation is to be founded is sufficient for the purpose, not available from other sources, and of a quality high enough to justify a fresh computation; and that (b) the result based on this material is so significantly different from previously published orbits as to warrant publication in the interest of progress in the knowledge of double stars and their behaviour.

Unless both are so, it would be wiser to await a later favourable opportunity.

2 Given the elements of the true orbit of a visual binary star, to compute the position angle θ, and separation ρ, at any epoch. (See also Programmes 15 to 18 in the Appendix.)

Introduction: It is recommended, when drawing up a programme for double-star observations, that a short selection of those stars which are closing up should be included, so that efforts may be made to make measures close to and through periastron. Also, when the elements of an orbit have been computed, it is advisable whenever possible to continue to record current measures for comparison with those computed from the elements. In this manner it is possible to derive differential corrections for the elements from the $C-O$ differences (computed minus observed positions), and thus, over a period of time, to make it possible to improve the computed orbit if the differences are significant.

The computation includes an approximate correction for the effect of precession on the position angle. No such correction is necessary for the distance. Unless the polar distance of the binary is small, or the interval of time from the standard epoch of the orbital elements is long, the correction to the position angle is very small—to the extent that for approximate work (say to an accuracy of $\pm 0°.5$) it can safely be ignored. Should this be the case, the computer may choose to skip Step 14 of Method 2A, or Step 6 of Method 2B.

In a case where the orbital elements are your own, naturally you will also know the standard epoch of the original measures (if this reduction has been necessary) and thus the epoch to which ω and Ω relate. As an extension of the computation of the elements it is highly desirable that you should also publish an ephemeris for at least 20 years into the future so that $C-O$ residuals can easily be obtained by other workers; Programmes 15 to 18 in Appendix II will make this task simple.

Whereas in the case of elliptical elements for comets the epoch for ω, Ω and i is always stated, this is not the case for binary stars. Usually only the date of publication of the orbit is quoted in readily accessible secondary sources of reference data such as the section of the *Atlas Cæli Catalogue* which gives the elements of double-star orbits. Observatory circulars sometimes give the epoch; many elements are endorsed 'precession ignored', or 'precession negligible'. In those cases where an epoch is given it is often 1900.0, because a large proportion of the measures used go back to 1850 or thereabouts; a few of the orbits published in the 1950s give the epoch as 1950.0. Nowadays, Heintz almost always works to the epoch 2000.0.

Apart from the fact that it is the task of the IAU Double-Star Commission, it is difficult to suggest a hard-and-fast rule for the guidance of computers of ephemerides where the elements are not their own, and where the epoch is unknown. As a rule-of-thumb I would suggest that the effect of precession on position angle should be ignored for all binaries within the declination range 25° N–25° S, and to assume for the remainder that the epoch is 1950.0 in the absence of any information to the contrary. As more orbits are revised in the light of later measures the position will no doubt regularize itself.

In practice, it will be found that any error introduced by adopting this assumption will be negligible, and one which is probably of the same order of magnitude as the uncertainty in the measurement of a current position angle of a close double. For example, in the 'further-practice' problems at the end of this topic, where 1950.0 is the assumed epoch, (a) and (d) give the same results as those published by Muller and Meyer in the *Troisième Catalogue d'Ephémérides d'Étoiles Doubles*, but (b) and (c) have an error of about 0°.3 in the position angle. However, if the epoch for (b) and (c) is taken as 1900.0 (see the Appendix) the results then agree with Muller and Meyer.

For the amateur, or the lone worker who is unable to research the required data from the publications of the observatories or the IAU, this question of epoch must remain something of a minor dilemma. It would certainly be a great help if, for instance, the compilers of such secondary data as the double-star information in the *Atlas Cæli Catalogue* were to include the epoch for ω and Ω. Perhaps, if space does not otherwise permit, this could be done by omitting the date of publication of the orbit. But if the computer adopts the above recommendation and assumes the epoch to be 1950.0, he may be reassured that any resulting error will be minimal for all practical purposes if the elements are, in fact, referred to epoch 1900.0 or 2000.0.

There is one comprehensive orbit catalogue which will be found to be invaluable. This is Finsen and Worley's *Third Catalogue of Orbits of Visual Binary Stars*, Rep. Obs. Circular No. 129, Johannesburg, 1970, and the interested reader should make every effort to consult this. An earlier catalogue was Worley's *Catalogue of Visual Binary Orbits*, Publications of the USNO, Second Series, Vol. XVIII, Part III,1963. The former is naturally more comprehensive and up-to-date. For the benefit of those who do not have access to either of these catalogues, and whose only reference is the *Atlas Cœli Catalogue*, I include in Appendix I a list of pairs, by ADS number, where the orbit is still current in Finsen and Worley's 1970 catalogue, and giving the epoch (if quoted) and the name of the computer.

The equations:

$$M = n(t - T) = E - e^{\circ} \sin E \qquad (7.28)$$

$$r = a(1 - e \cos E) \qquad (7.29)$$

$$\tan\tfrac{1}{2}v = \sqrt{\frac{1+e}{1-e}} \times \tan\tfrac{1}{2}E \qquad (7.30)$$

$$\tan(\theta - \Omega) = \tan(v + \omega)\cos i \qquad (7.31)$$

$$\rho = r\cos(v + \omega)\sec(\theta - \Omega) \qquad (7.32)$$

$$\Delta\theta = +0^{\circ}.005\,6 \sin\alpha \sec\delta\,(t - t_0) \qquad (7.33)$$

where:

M = the mean anomaly, expressed in degrees
n = the mean annual motion, in degrees
t = the desired epoch of polar coordinates
E = the eccentric anomaly, in degrees
T, e, a, ω, Ω, and i are the orbital elements as defined in Topic 1
r = the radius vector
v = the true anomaly, in degrees
θ = the position angle, in degrees
ρ = the separation, in seconds of arc
t_0 = the standard epoch, where necessary

$$e^{\circ} = \frac{180\,e}{\pi}$$

Further information: R. G. Aitken, *The Binary Stars*, pp 79–80; for differential corrections, pp 109–113; for effect of precession on position angle, p 73 and *Astronomy—A Handbook*, Ed. G. D. Roth, pp 474 and 485. Bate, Mueller and White, *Fundamentals of Astrodynamics*, Chap. 4; for effect of proper motion on position angle, P. van de Kamp, *Principles of Astrometry*, pp 25–26 and p 144.

Example 2. Find the position angle θ, and separation ρ, of ϵ^1 Lyr ($18^h\,42^m\,40^s.87$, $+39^{\circ}\,37'\,00''$) for 1978.0, given the following orbital elements (Güntzel-Lingner, 1954, *Atlas Cœli Catalogue*), assuming the epoch for ω and Ω to be 1950.0:

$P = 1\,165.6$ years
$T = 2\,318$
$e = 0.19$
$a = 2''.78$
$\omega = 165°.7$
$\Omega = 29°$
$i = 138°$
$n = \dfrac{360}{P} = 0°.308\,85$

and $e° = 10°.886\,2$

Note: Eqn. 7.28 includes the expression $n\,(t - T)$. In this example T is in the future, so obtain the time of last periastron passage from $T - P = 1\,152.4$. Before commencing the calculation proper, iterate for E in Eqn. 7.28 where, after substituting, we have:

$$0.308\,85 \times 825.6 = E - 10.886\,2\,(\sin E)$$
$$254.99 = E - 10.886\,2\,(\sin E)$$
$$E = 245°.11.$$

This is the value for E we shall use in the two calculator methods to be demonstrated.

But first, those who dislike having to iterate for E might pose the question, 'What about the Equation of the Centre? This is:

$$v = M + (2e - \tfrac{1}{4}e^3)\sin M + \frac{5}{4}e^2\sin 2M + \frac{13}{12}e^3\sin 3M, \tag{7.34}$$

which enables v to be found directly in terms of e (not $e°$) and M (expressed in radians), without iterating for E in Eqn. 7.28. Remember that the lefthand side of Eqn. 7.28 will give M in radians if n is expressed in radians, so v will also be given in radians by the Equation of the Centre. When v is converted into degrees, users of Method 2B can proceed directly from Step 4.

Certainly time will be saved, but caution is advised—the Equation of the Centre is derived from a series expansion, normally truncated after the term in e^3, and it is only suitable for low values of the eccentricity ($e < 0.2$). You will find, in fact, that if this method is used, attractive though it seems, the small error due to truncation will lead to an error of about 0°.1 in v and in the position angle (stored in R_2 at the end of Step 4, Method 2B). In the worked example, θ thus derived is 355°.82, compared with 355°.70 from the working demonstrated.

Having explored this diversion, let us proceed to work the example.

Method 2A.

1. Clear memory; enter E in degrees	**MC** **[245.11]** **f cos**
2. Multiply by e	× **[0.19]** =
3. Change sign; add 1	**CS** + **1**

Method 2A *continued*

Step	Keys	Result
4. Multiply by a	×	
	[2.78]	
5. Note r for Step 13	=	$r = 3.002$
6. Enter e	**[0.19]**	
	M+	
	CS	
	+	
	1	
	=	
	←→	
	XM	
	+	
	1	
	÷	
	MR	
	=	
	f $\sqrt{x}$	
	MC	
	M+	
6. Enter E	**[245.11]**	
	÷	
	2	
	=	
	f tan	
	×	
	MR	
	=	
	f $\tan^{-1}$	
	STOP	
7. Examine displayed $\frac{1}{2}v$: if negative, add 180°; if positive, skip operation in brackets	[+ 180]	(–62.623 06) Display negative: operation included
8. Multiply by 2 for v	×	
	2	
Note v for Step 13	=	$v = 235°.553$
9. Add ω	+	
	[165.7]	
	=	
	f tan	
	MC	
	M+	
10. Enter i	**[138.0]**	
	f cos	
	×	
	MR	
	=	
	f $\tan^{-1}$	
11. Add Ω	+	
	[29.0]	
	MC	
	M+	
	=	
	STOP	(–4.096)

Method 2A *continued*

If displayed θ is negative, add 360°; if positive, skip operation in brackets	[+ 360 =]	θ negative: operation carried out
Note θ		θ (uncorrected) = 355°.904
12. Evaluate ρ	– **MR** = **f cos** **f** $\frac{1}{x}$ **MC**	
Store $\sec(\theta - \Omega)$	**M+**	
13. Enter ν from Step 8	**[235.553]**	
Add ω	+ **[165.7]** = **f cos** × **MR** ×	
Enter r from Step 4	**[3.002]**	
Note ρ: if display is negative, change sign; also add 180° to θ. If θ now > 360°, deduct 360°	=	$\rho = 2''.69$
14. If required, correct θ for precession; if not, computation is now complete		
Enter α in decimal hours; convert to degrees	**[18.711 361]** × **15** = **f sin** × **0.005 6** = **MC** **M+**	
Enter δ in decimal degrees	**[39.617 5]** **f cos** **f** $\frac{1}{x}$ × **MR** = **MC** **M+**	
Enter year for which position required	**[1978]** –	

Method 2A *continued*

Enter assumed standard epoch	**1950**	Unless the epoch is known to be 1900.0 or 2000.0
	×	
	MR	
	+	
Enter θ from Step 11 (or adjusted θ from Step 13)	**[355.904]**	
Note θ, corrected for precession	=	$\theta = 355°.70$

Result 2A. Based on the orbital elements of Güntzel-Lingner (1954) from the *Atlas Cæli Catalogue*, the position angle and separation of ϵ^1 Lyr for 1978.0 are 355°.7 and 2″.69 respectively. Muller and Meyer, *Troisième Catalogue d'Ephémérides d'Étoiles Doubles*, quote the same values. Check of the USNO *Catalogue of Visual Binary Orbits* shows that No. 405, ϵ^1 Lyr, has 1950.0 for the standard epoch, so the assumption of the epoch is justified in this case.

Method 2B.

1. Fix 2 decimal places	**f FIX 2**	
Store: *E*	**[245.11]**	
	STO 0	
e	**[0.19]**	
	STO 1	
a	**[2.78]**	
	↑	'Enter'
2. Evaluate and store *r*	**1**	
	RCL 1	
	RCL 0	
	f cos	
	×	
	−	
	×	
	RCL 0	
	$x \longleftrightarrow y$	
	STO 0	(r in R_0)
	R ↓	
3. Evaluate and store *v*	**2**	
	÷	
	f tan	
	1	
	RCL 1	
	+	
	1	
	RCL 1	
	−	
	÷	
	f $\sqrt{x}$	
	×	
	g $\tan^{-1}$	
	STOP	
If displayed $\frac{1}{2}v$ is negative, add 180°; if not, skip operation in brackets	[**180** +]	(−62.22) Display negative: operation carried out
	2	
	×	

Method 2B *continued*

4. Add ω	[165.7]	
	+	
	STO 1	($\nu + \omega$ in R_1)
	f tan	
Enter i	[138.0]	
	f cos	
	×	
	g tan^{-1}	
Enter Ω	[29.0]	
	+	
	f last x	
	$x \longleftrightarrow y$	
	STOP	
If displayed θ is negative, add 360°; if not, skip operation in brackets	[360 +]	Display negative: operation carried out
5. Evaluate ρ	**STO 2**	(Uncorrected θ)
	$x \longleftrightarrow y$	
	−	
	f cos	
	g $\frac{1}{x}$	
	RCL 1	
	f cos	
	×	
	RCL 0	
	×	
	STO 0	(ρ in R_0)
6.(a) If θ is not to be corrected for precession, the computation is now complete; if a corrected θ is required, skip Step 6(a)		
Read θ	**RCL 2**	Uncorrected θ = 355°.90
Read ρ	$x \longleftrightarrow y$	ρ = 2″.69
(If ρ display is negative, carry out the operations in brackets and read θ in correct quadrant)	[**CHS** $x \longleftrightarrow y$ **180** +]	(ρ is positive, so operations not carried out)
(b) Correct θ for precession		
Enter α in H.MS format and convert to degrees	[18.42 40 9]	
	g → H	
	15	
	×	
	f sin	
Enter δ in D.MS format	[39.37 03]	CHS if Southern dec.
	g → H	
	f cos	
	g $\frac{1}{x}$	
	×	
Enter t, year for which position angle is required	[1978]	
	↑	

Method 2B *continued*

Enter standard epoch t_0 for the elements	**1950**	(Unless it is known to be 1900.0 or 2000.0)
	−	
	×	
	0.005 6	
	×	
	RCL 2	
Read corrected θ	+	$\theta = 355°.70$
Read ρ (If ρ display is negative, carry out the operations in brackets and read θ in correct quadrant. If θ now > 360°, deduct 360°)	**RCL 0**	$\rho = 2''.69$ (ρ is positive, so operations not carried out)
	[**CHS**	
	$x \longleftrightarrow y$	
	180	
	+]	

Result 2B. The RPN method gives the same result as Method 2A, $\theta = 355°.7$, $\rho = 2''.69$. Muller and Meyer, *Troisième Catalogue d'Ephémérides d'Étoiles Doubles*, quote the same values. See the remarks with Result 2A regarding the assumption of 1950.0 for epoch.

Proper motion: References in the literature about the effect of proper motion on position angle may cause surprise that no mention has been made of it so far in this Chapter. Only the proper motion in RA is relevant. Where this is unusually large and the NPD is small, a further correction to the position angle is justified if over the period $t - t_0$ it would amount to as much as $\pm 0°.05$. The equation is:

$$\Delta\theta_2 = +0°.004\,17\ \mu_\alpha \sin\delta\ (t - t_0),$$

where μ_α is expressed in seconds of time.

If necessary, this additional correction is applied at the end of the computation, after the correction for the effect of precession. It will be found necessary only very rarely.

Whatever the declination of the primary, $\Delta\theta_2$ will be less than $\pm 0°.05$ (and thus can be ignored) for any period up to 25 years, if the annual proper motion in RA is less than $\pm 0^s.479\,1$. Of the nearest stars which are also binaries, α Cen has a proper motion μ_α of $-0^s.490\,4$, 61 Cyg has $+0^s.352\,3$, Gr 34 has $+0^s.265\,0$, Σ2398 has $-0^s.178\,9$. Of these, only α Cen approaches $0^s.5$ per annum. Even at the pole, if μ_α is as much as $\pm 0^s.5$, $\Delta\theta_2$ will only exceed $\pm 0°.05$ if $t - t_0 > 24$ years. Where $\delta < 50°$ and $\mu_\alpha = \pm 0^s.5$, $\Delta\theta_2$ will only exceed $\pm 0°.05$ if $t - t_0 > 31$ years.

As an example, take the case of α Cen, using Heintz's elements for 2000.0 and compute θ, ρ for 1975.0, correcting for precession. The result is $\theta = 207°.24$, $\rho = 20''.92$, agreeing with Muller and Meyer in the *Troisième Catalogue*. If θ is now corrected for the effect of proper motion, the amount of the correction is only $-0°.04$. Thus θ remains unchanged at $207°.2$ when rounded to the single decimal place.

So, it is obvious that it will be very rare for a correction for the effect of the proper motion to be justified, over the time-scales we normally encounter.

The correction will only need to be considered for

α Cen if $t - t_0 > 27.5$ years
61 Cyg if $t - t_0 > 54$ years
Gr 34 if $t - t_0 > 64$ years
$\sum$ 2398 if $t - t_0 > 76$ years

and still might not have a significant effect. Should θ be, say, 53°.33 and the correction required be –0°.06, θ will remained unchanged at 53°.3 when rounded to one decimal place. But if it is 53°.28 and the correction is applied, then θ becomes 53°.2 when rounded.

For further practice, try the following:

(a) Find θ, ρ for ϵ^2 Lyr (*ADS* 11 635), $18^h 42^m 43^s.33$, +39° 33′ 34″, at 1975.0, given the orbital elements (Güntzel-Lingner, 1954) from the *Atlas Cæli Catalogue*: $P = 585$ years, $T = 2229.5$, $e = 0.49$, $a = 2''.95$, $\omega = 92°.0$, $\Omega = 17°.4$, $i = 120°.5$. Correct θ for precession, assuming the epoch for the elements to be 1950.0.

(b) Find the position angle and separation of α Gem (*ADS* 6 175), $7^h 31^m 24^s.65$, +31° 59′ 59″, at 1978.0, given the orbital elements (Rabe, 1957): $P = 420.07$, $T = 1965.30$, $e = 0.33$, $a = 6''.29$, $\omega = 261°.4$, $\Omega = 40°.5$, $i = 115°.9$. Correct θ for precession, assuming the epoch for the elements to be 1950.0.

(c) Find θ, ρ for γ Lup (*h*. 4 786), $15^h 31^m 47^s.99$, –41° 00′ 01″, at 1977.0, given the orbital elements (Heintz, 1956): $P = 147$ years, $T = 1887.0$, $e = 0.49$, $a = 0''.59$, $\omega = 301°$, $\Omega = 92°.8$, $i = 95°.6$. Correct θ for precession, assuming the epoch for the elements to be 1950.0.

(d) Find the position angle and separation of $\sum$ 3062 (*ADS* 61), $0^h 03^m 38^s.19$, +58° 09′ 29″, for 1977.0, given the orbital elements (Baize, 1957): $P = 106.83$ years, $T = 1943.05$, $e = 0.45$, $a = 1''.43$, $\omega = 98°.8$, $\Omega = 39°.1$, $i = -44°.4$. Correct θ for precession, assuming the epoch for the elements to be 1950.0.

(e) Using apertures of 40 and 82 inches, van Biesbroeck measured 24 Aqr from 1935 to 1943 (Pub. Yerkes Obs., VIII, Pt. VI).

After listing the separate measures, van Biesbroeck quotes these averages:

1937.01	250°.0	0″.38	3 *n*
1941.16	261°.1	0″.41	4 *n*
1943.72	263°.9	0″.48	5 *n*

Duruy, with a 40-cm reflector, made two measures of 24 Aquarii:

1965.80	275°	0″.45	3 *n* (θ)	1 *n* (ρ)
1967.68	306°	0″.35	5 *n*	

Assuming these measures to be typically reliable, calculate the $C - O$ differences, using:

(*i*) the orbital elements calculated in Topic 1;
(*ii*) Danjon's orbital elements (listed for comparison at the end of the computation in Topic 1).

In view of the declination of 24 Aquarii, there is no need to correct θ for precession.

Your results should be:

(a) $\theta = 85°.8$, $\rho = 2''.32$. Muller and Meyer quote the same values. If your result differs from this, check that your value for E was computed to be 195°.76.

(b) $\theta = 100°.9$, $\rho = 2''.10$. Muller and Meyer quote $\theta = 101°.2$, $\rho = 2''.10$. If your

result differs from the one given, check that your value for E was computed to be $16^\circ.15$. This is a case where the assumption for epoch leads to a small error in the computed position angle ($-0^\circ.3$). Rabe's epoch for ω and Ω is actually 1900.0, and if this is employed in the correction for precession instead of the assumed 1950.0, the result agrees exactly with Muller and Meyer.
(c) $\theta = 277^\circ.8$, $\rho = 0''.62$ (after changing the sign for ρ and adding 180° to θ). Muller and Meyer quote $\theta = 277^\circ.5$, $\rho = 0''.62$. If your result differs from the one given, check that your value for E was computed to be $207^\circ.46$. Again, Heintz's epoch was 1900.0, and if this is used instead of the assumed epoch of 1950.0, the result agrees with Muller and Meyer.
(d) $\theta = 280^\circ.8$, $\rho = 1''.42$ (after changing the sign for ρ and adding 180° to θ). Muller and Meyer quote the same values. If your result differs from this, check that your value for E was computed to be $133^\circ.20$.
(e) The residuals are:

		$C-O$			
Observer	Epoch	Danjon		Topic 1	
		θ	ρ	θ	ρ
van Biesbroeck	1937.01	$+1^\circ.32$	$+0''.03$	$-1^\circ.16$	$+0''.11$
,,	1941.16	$-1^\circ.59$	$+0''.06$	$-5^\circ.08$	$+0''.13$
,,	1943.72	$-0^\circ.19$	$+0''.01$	$-3^\circ.97$	$+0''.07$
Duruy	1965.80	$+22^\circ.45$	$-0''.07$	$+19^\circ.63$	$+0''.01$
,,	1967.68	$-2^\circ.79$	$-0''.03$	$-6^\circ.99$	$+0''.08$

Reference to many more reliable modern measures would be required before any sensible conclusions could be drawn about the revisions necessary to improve the orbital elements, but it is obvious that Danjon's orbit is superior to the one we calculated in Topic 1, on this small sample. Check, if necessary, your values for E against:

	for Danjon	for Topic 1
1937.01	$137^\circ.06$	$117^\circ.64$
1941.16	$155^\circ.0$	$136^\circ.46$
1943.72	$165^\circ.46$	$146^\circ.92$
1965.80	$267^\circ.10$	$231^\circ.89$
1967.68	$282^\circ.01$	$240^\circ.65$

NOTES

8 Ephemerides of Comets

Topic 1 Finding the equatorial coordinates α, δ, at any date, for a newly-discovered comet, given the preliminary parabolic elements.

Topic 2 Finding the equatorial coordinates α, δ, at any date, for a recovered periodic comet, given the elliptical elements.

Preliminary remarks: In the previous Chapter we solved problems involving the apparent motion of one body around another in an elliptical orbit. Having obtained a set of elements defining the orbit, we can predict simply the position of the companion body at any time, relative to the primary. However, this computation of orbital position, based on Kepler's laws of motion, is valid only for the case where the two bodies are alone in space and not subject to the gravitational influences of other bodies. Where more than one body is involved in orbital motion about the primary, disturbances will sometimes occur which will alter the orbits. The effect is a varying one, depending upon the distance between and the relative masses of the orbiting bodies; consequently, the determination of the extent of such gravitational perturbations can become an involved process, as it is very unlikely that the two bodies will have similar periods or masses.

In many instances the effect of the perturbative forces will be so small that they can be ignored; on the other hand a short-period comet, for example, may follow a regular and undisturbed elliptical orbit around the Sun for several revolutions and then (probably at aphelion) come into conjunction with Jupiter (Sun, comet and Jupiter all lying on or very close to the same line), so that the close proximity of the mass of the giant planet causes an acceleration in the heliocentric velocity of the comet. The total velocity may be sufficient to change the orbit from an ellipse to a hyperbola. In such an event the comet would be carried out of the Solar System altogether, never to return. Another effect of the perturbative forces might be to split a comet up into two or more parts; even, eventually, to spread such debris out along the orbit and to give rise to a meteoroid stream.

Reports of discoveries of new comets sent to national or international coordinating bodies are published immediately, so that a mass of observational data will enable the orbital elements to be quickly derived. The observations will naturally cover only a very small part of the trajectory, so that it may, initially at least, be treated as a parabola. In this case, the preliminary orbit will usually be described in terms of parabolic elements (see Topic 1), unless it is reasonably certain that the new comet is a previously unknown (or lost) one of short period, when elliptical elements will probably be computed.

Predictions for the recovery of a periodic comet will, of course, be given in terms of elliptical elements, and the calculated effects of perturbation by the planetary masses of the Solar System will be incorporated in the predictions.

Although there are many similarities between the computation required and that given in the last chapter for the two-body problem, it is my view that it is far better to leave the computation of the orbital elements of comets, and of the perturbative effects of the planets, in the hands of recognized experts. In this Chapter, the topics are confined to computing positions in the orbit from previously published elements, parabolic or elliptical, and converting those positions into geocentric equatorial coordinates. Complete programmes for the HP-67 calculator are included in Appendix II.

Further information: There is a wealth of literature devoted to the subject of comets, their orbits, and perturbations. The subject demands, and gets, book-length treatment. The reader seeking further information is recommended initially to consult: *Sky and Telescope*, April 1977, p 306 et seq; J. B. Sidgwick, *Observational Astronomy for Amateurs*, Section 16; *Astronomy—A Handbook*, ed. G. D. Roth, Chap. 16, and Table 17 in the Appendix (orbital elements for periodic comets with periods under 200 years); R. M. L. Baker and M. W. Makemson, *An Introduction to Astrodynamics*, 1.7 (perturbations) and Chap. 3; R. M. L. Baker, *Astrodynamics*, Chaps. 1–4 (advanced treatment of orbit determination, improvements and perturbations); *Smithsonian Catalogue of Cometary Orbits*, 2nd edition; W. M. Smart, *Spherical Astronomy*, Chap. 5 (planetary motions). The reader who wishes to specialize in this field can find a definitive treatment, with fully-worked examples of orbit computation for comets and minor planets, and perturbations, in A. D. Dubyago, *The Determination of Orbits*.

1 To find the equatorial coordinates, α, δ, at any date, for a newly-discovered comet, given the preliminary parabolic elements T, ω, Ω, i, q

where T = the time of perihelion passage

ω = the angle in the plane of the orbit between the node and the point of perihelion passage

Ω = the longitude of the ascending node, measured in the plane of the ecliptic

i = the inclination of the orbit, that is, the angle between the plane containing the orbit and that of the ecliptic. If the motion is direct (anticlockwise as seen from the North pole of the ecliptic) i lies between 0° and 90°; if retrograde, between 90° and 180°

q = the perihelion distance, expressed in AU

The equations:

$$\tan\frac{v}{2} + \frac{1}{3}\tan^3\frac{v}{2} = 0.012\,163\,7\,\frac{t}{q^{3/2}} \tag{8.1}$$

If you are using a 10-digit calculator, the constant is 0.012 163 721.

$$r = q \sec^2 \frac{v}{2} \tag{8.2}$$

$$\left.\begin{aligned} x_1 &= r(\cos\Omega \cos(v + \omega) - \sin\Omega \sin(v + \omega) \cos i) \\ y_1 &= r(\sin\Omega \cos(v + \omega) + \cos\Omega \sin(v + \omega) \cos i) \\ z_1 &= r \sin(v + \omega) \sin i \end{aligned}\right\} \tag{8.3}$$

$$\left.\begin{aligned} x &= x_1 \\ y &= y_1 \cos\epsilon - z_1 \sin\epsilon \\ z &= y_1 \sin\epsilon + z_1 \cos\epsilon \end{aligned}\right\} \tag{8.4}$$

$$\left.\begin{aligned} \xi &= x + X \\ \eta &= y + Y \\ \zeta &= z + Z \end{aligned}\right\} \tag{8.5}$$

$$\left.\begin{aligned} \tan\alpha &= \frac{\eta}{\xi} \\ \Delta \cos\delta &= \frac{\xi}{\cos\alpha} = \frac{\eta}{\sin\alpha} \\ \tan\delta &= \frac{\zeta}{\Delta \cos\delta} \\ \Delta &= (\Delta \cos\delta) \sec\delta \end{aligned}\right\} \tag{8.6}$$

where $v =$ the true anomaly, positive after perihelion, negative before

$t =$ the time interval in days between the time for which α and δ are required and T, positive after perihelion, negative before

$q =$ the perihelion distance, in AU

$r =$ the radius vector from the centre of the Sun, at time t, in AU

$x_1, y_1, z_1, =$ the heliocentric ecliptic rectangular coordinates of the comet at time t

$x, y, z =$ the heliocentric equatorial rectangular coordinates of the comet at time t

$\epsilon =$ the obliquity of the ecliptic at the same epoch as the mean equator and equinox for ω, Ω and i (usually 1950.0)

$X, Y, Z =$ the geocentric equatorial rectangular coordinates of the Sun at time t, referred to the same mean equator and equinox (usually 1950.0)

$\xi, \eta, \zeta =$ the geocentric equatorial rectangular coordinates of the comet at time t, referred to the same epoch as X, Y, Z

$\alpha, \delta =$ the RA and dec. of the comet, referred to the same epoch as X, Y, Z

$\Delta =$ the distance of the comet from the centre of the Earth, in AU, at time t

It is clear that v, to be determined as a function of the time and perihelic distance, being transcendental, is not so easy to compute as in the case of an elliptical orbit where e and a are known and where r can be obtained after solution of Kepler's equation for the eccentric anomaly E (Eqns. 7.28 and 7.29). In the absence of a reference table for $v + \frac{1}{3}v^3$, where $v = \frac{\tan v}{2}$, an iterative solution for v must be found. Users of HP-25 calculators can make use of Programme 20 in the Appendix. Users of other calculators will have no difficulty in producing a converging result in a very few steps. (HP-67 users will prefer to use the complete programme in the Appendix, Programme 21.)

Example 1. The following provisional parabolic elements for Comet Bradfield 1975p were published by the BAA on 1975, December 3. Compute α, δ, Δ and r for 0^h ET on 1976, February 2.

$T =$ 1975, December 21.173 1 ET
$q =$ 0.218 445 AU
$\omega =$ 358°.129 0
$\Omega =$ 270°.625 7
$i =$ 70°.635 7

Epoch for ω, Ω and $i =$ 1950.0.

Before proceeding to the computation proper, iterate for v in Eqn. 8.1 (after evaluating the right-hand term as 5.102 337) and obtain $v =$ 128°.737 8.

Method 1A.

1. Find r:		
Enter v	**[128.737 8]**	
	÷	
	2	
	=	
	f cos	
	f $\frac{1}{x}$	
	×	
	=	
	×	
Enter q	**[0.218 445]**	
Note r	**=**	$r = 1.167$
2. Find remaining terms for x_1, etc.:		
Enter Ω	**[270.625 7]**	
	M+	
Note A	**f cos**	$\cos\Omega = 0.010\,921 = A$
	MR	
Note B	**f sin**	$\sin\Omega = -0.999\,94 = B$
Enter v	**[128.737 8]**	
	+	
Add ω	**[358.129 0]**	
	=	
	MC	
	M+	
Note C	**f cos**	$\cos(v + \omega) = -0.599\,956 = C$
	MR	
Note D	**f sin**	$\sin(v + \omega) = 0.800\,033 = D$
Enter i	**[70.635 7]**	
	MC	
	M+	
Note E	**f cos**	$\cos i = 0.331\,574 = E$
	MR	

Method 1A *continued*

Note F	f sin	$\sin i = 0.943\ 429 = F$
3. Find x_1:		
Enter B	[0.999 94]	
	[CS]	B is negative
	×	
Enter D	[0.800 033]	
	×	
Enter E	[0.331 574]	
	=	
	MC	
	M+	
Enter A	[0.010 921]	
	×	
Enter C	[0.599 956]	
	[CS]	C is negative
	−	
	MR	
	×	
Enter r	[1.167]	
Note x_1	=	$x_1 = 0.301\ 905\ 2$
4. Find y_1:		
Enter B	[0.999 94]	
	[CS]	B is negative
	×	
Enter C	[0.599 956]	
	[CS]	C is negative
	=	
	MC	
	M+	
Enter A	[0.010 921]	
	×	
Enter D	[0.800 033]	
	×	
Enter E	[0.331 574]	
	+	
	MR	
	×	
Enter r	[1.167]	
Note y_1	=	$y_1 = 0.703\ 487\ 3$
5. Find z_1:		
Enter D	[0.800 033]	
	×	
Enter F	[0.943 429]	
	×	
Enter r	[1.167]	
Note z_1	=	$z_1 = 0.880\ 821\ 6$
6. Note x (= x_1 from Step 3)	[No operation]	$x = 0.301\ 905\ 2$
7. Enter ϵ_{1950} in degrees	23.445 788	
	MC	
	M+	
Note G	f cos	$\cos\epsilon = 0.917\ 437 = G$
	←→	
	XM	
Note H	f sin	$\sin\epsilon = 0.397\ 881 = H$

Method 1A *continued*

8. Find y:		
Enter y_1 (Step 4)	**[0.703 487 3]**	
	×	
	MR	
	=	
	MC	
	M+	
Enter z_1 (Step 5)	**[0.880 821 6]**	
	×	
Enter H	**[0.397 881]**	
	=	
	←→	
	XM	
	−	
	MR	
Note y	=	$y = 0.294\ 943\ 1$
9. Find z:		
Enter y_1 (Step 4)	**[0.703 487 3]**	
	×	
Enter H	**[0.397 881]**	
	=	
	MC	
	M+	
Enter z_1 (Step 5)	**[0.880 821 6]**	
	×	
Enter G	**[0.917 437]**	
	+	
	MR	
Note z	=	$z = 1.088\ 002\ 5$
10. Find ξ:		
Enter x (Step 6)	**[0.301 905 2]**	
	+	
Enter X_{1950}	**[0.658 151 9]**	
Note ξ	=	$\xi = 0.960\ 057\ 1$
11. Find η:		
Enter y (Step 8)	**[0.294 943 1]**	
	+	
Enter Y_{1950}	**[0.672 908 4]**	
	[CS]	Y is negative
Note η	=	$\eta = -0.377\ 965\ 3$
12. Find ζ:		
Enter z (Step 9)	**[1.088 002 5]**	
	+	
Enter Z_{1950}	**[0.291 786 3]**	
	[CS]	Z is negative
Note ζ	=	$\zeta = 0.796\ 216\ 2$
13. Find α:		
Enter η	**[0.377 965 3]**	
	[CS]	η is negative
	÷	
Enter ξ	**[0.960 057 1]**	
	=	
	f $\tan^{-1}$	
	÷	
	15	

Method 1A *continued*

	=	
	STOP	(–1.432 606)

If $\eta+$, $\xi+$, α is between 0^h and 6^h; if $\eta+$, $\xi-$, α is between 6^h and 12^h; if $\eta-$, $\xi-$, α is between 12^h and 18^h; if $\eta-$, $\xi+$, α is between 18^h and 24^h. If necessary adjust the display at this stage by multiples of 6^h until α appears in the correct quadrant.

Note α in hours	[+ 24 =]	Adjustment made. $22^h.567\,394$
Deduct integral hours; convert to minutes	–	
	[22]	
	×	
	60	
Note minutes	=	$34^m.044$ $\alpha = 22^h\,34^m.04$
14. Find and store $\Delta\cos\delta$:		
Enter ξ	**[0.960 057 1]**	
	MC	
	M+	
Enter α in decimal hours	**[22.567 394]**	
	×	
	15	
	=	
	f cos	
	←→	
	XM	
	÷	
	MR	
	=	
	MC	
	M+	
15. Find δ:	**[0.796 216 2]**	
Enter ζ	÷	
	MR	
	=	
Note δ in degrees	**f tan^{-1}**	$\delta = 37^\circ.657\,1$
Deduct integral degrees;	–	
convert to minutes	[37]	CS if Southern dec.
	×	
	60	
Note minutes	=	$39'.426$ $\delta = +37^\circ\,39'.4$
16. Find Δ:		
Enter δ in degrees	**[37.657 1]**	
	f cos	
	f $\frac{1}{x}$	
	×	
	MR	
Note Δ	=	$\Delta = 1.303$

Result 1A. The equatorial coordinates for Comet Bradfield 1975*p* at 0^h ET on

1976, February 2 were $\alpha = 22^h 34^m.04$, $\delta = +37° 39'.4$. The radius vector from the centre of the Sun, r, was 1.167 and the distance of the comet from the centre of the Earth was 1.303 AU. As a check, the coordinates published in *BAA Circular No.* 570 were $\alpha = 22^h 34^m.07$, $\delta = +37° 40'.1$, $r = 1.167$, $\Delta = 1.304$. There are some small differences in the computation: in α, $-0^m.03$; in δ, $-0'.7$; in r, nil; in Δ, $+0.001$ AU. The accuracy is good enough to place the comet near the centre of the field of the telescope, provided that the orbital elements were based on a sufficient number of accurate early observations.

Method 1B.

1. Find r:		
Enter v	**[128.737 8]**	
	STO 7	
	2	
	÷	
	f cos	
	g $\frac{1}{x}$	
	g x^2	
Enter q	**[0.218 445]**	
	×	
Note r	**STO 6**	$r = 1.167$
2. Evaluate terms of Eqn. 8.3:		
Enter Ω	**[270.625 7]**	
	f cos	
	STO 0	
	f last x	
	f sin	
	STO 1	
Enter ω	**[358.129 0]**	
	RCL 7	
	+	
	f cos	
	STO 2	
	f last x	
	f sin	
	STO 3	
Enter i	**[70.635 7]**	
	f cos	
	STO 4	
	f last x	
	f sin	
	STO 5	
Enter ϵ_{1950}	**23.445 788**	
	STO 7	
3. Find α	**RCL 0**	
	RCL 2	
	×	
	RCL 1	
	RCL 3	
	×	
	RCL 4	
	×	

Method 1B *continued*

−
RCL 6
×
RCL 1
RCL 2
×
RCL 0
RCL 3
×
RCL 4
×
+
RCL 6
×
STO 1
$x \longleftrightarrow y$
STO 0
RCL 6
RCL 3
RCL 5
×
×
STO 2
RCL 1
RCL 7
f cos
×
RCL 2
RCL 7
f sin
×
−
STO 3
RCL 1
RCL 7
f sin
×
RCL 2
RCL 7
f cos
×
+
STO 4
RCL 3
STO 1
RCL 4
STO 2

Enter X_{1950}	[0.658 151 9	
	STO + 0	
Enter Y_{1590}	[0.672 908 4]	
	[CHS]	Y is negative
	STO + 1	
Enter Z_{1950}	[0.291 786 3]	
	[CHS]	Z is negative

Method 1B *continued*

	STO + 2	
	RCL 1	
	RCL 0	
	g → P	
	R ↓	
	15	
	÷	
	STOP	(−1.432 116 978)
If display negative, add 24	[24 +]	Adjustment made
	STO 7	
Note α	f H.MS	$\alpha = 22^h\ 34^m\ 04^s.37$
4. Find δ:	RCL 0	
	RCL 7	
	15	
	×	
	f cos	
	÷	
	RCL 2	
	$x \leftrightarrow y$	
	STO 6	
	÷	
	g $\tan^{-1}$	
	STO 5	
Note δ	f H.MS	$\delta = +37°\ 40'\ 06''.44$
5. Find Δ:	RCL 5	
	f cos	
	g $\frac{1}{x}$	
	RCL 6	
Note Δ	×	$\Delta = 1.303\ 5$

Result 1B. The equatorial coordinates of Comet Bradfield 1975*p* at 0^h ET on 1976 February 2 were $\alpha = 22^h\ 34m\ 04^s.37$, $(22^h\ 34^m.07)$, $\delta = +37°\ 40'\ 06''.44$ $(+37°\ 40'.1)$. The radius vector from the centre of the Sun, r, was 1.167 and the distance of the comet from the centre of the Earth, Δ, was 1.304 AU. As a check, the co-ordinates published in *BAA Circular No. 570* were exactly the same. Thus the accuracy is marginally better than that achieved by Method 1A with a simple calculator.

Practice examples are included at the end of the Chapter.

2 To find the equatorial coordinates α, δ, at any date, for a newly-recovered periodic comet, given the elliptical elements of the orbit *T*, *P*, *e*, *a*, *n*°, ω, Ω, *i* and *q*

where T = the time of perihelion passage
P = the period of the comet, in years
e = the numerical eccentricity of the orbit
a = the semi-major axis, expressed in AU

$n° =$ the mean daily motion, in degrees
$\omega =$ the angle in the plane of the orbit between the ascending node and the point of perihelion passage
$\Omega =$ the longitude of the ascending node, measured in the plane of the ecliptic
$i =$ the inclination of the orbit, that is, the angle between the plane containing the orbit and that of the ecliptic. If the motion of the comet is direct (anticlockwise as seen from the North pole of the ecliptic) i lies between 0° and 90°; if retrograde, between 90° and 180°
$q =$ the perihelion distance, expressed in AU

The equations:

$$M = n^\circ t = E - e^\circ \sin E \tag{8.7}$$

$$e^\circ = \frac{180\,e}{\pi} \tag{8.8}$$

plus Eqns. 7.29, 7.30 and 8.3 to 8.6,
where $M =$ the mean anomaly
$t =$ the time interval in days between T and the time for which the position is required; t is negative before T, positive after
$E =$ the eccentric anomaly, expressed in degrees

It will be apparent that in Eqn. 8.7 E is transcendental and can be found only by iteration. The technique is discussed in Step 5 of Topic 1, Chapter 7. Users of HP-25 calculators can, if they wish, use the iteration programme (Programme 19) in the Appendix, but the iteration is reasonably short and it is not necessary to use the programme. Users of other calculators will be able to produce a converging result in a very few steps. (Users of the HP-67 will prefer to use the complete programme in Appendix II, Programme 22.)

Example 2. Comet Smirnova-Chernykh 1975*e* is visible all around its orbit. *IAUC 2918* provides the following elliptical elements. Find α, δ, Δ and r for 1977, January 17, at 0^h ET.

$T =$ 1975, August 6.474 2 ET
$\omega = 90^\circ.219\,5$ (1950.0)
$\Omega = 77^\circ.102\,4$ (1950.0)
$i = 6^\circ.641\,3$ (1950.0)
$e = 0.145\,446$
$a = 4.174\,405$ AU
$n^\circ = 0.115\,561\,2$
$P = 8.529$ years
$q = 3.567\,253$ AU
(q is not required for the computation.)

Before proceeding to the computation proper, evaluate $M = n^\circ t$ and iterate for E in Eqn. 8.7, finding $M = 61^\circ.192\,637$ and $E = 68^\circ.971\,1$.

Method 2A.

---

1. Find r:		
Enter E in degrees	**[68.971 1]**	
	f cos	
	×	
Enter e	**[0.145 446]**	
	M+	
	CS	
	+	
	1	
	×	
Enter a	**[4.174 405]**	
Note r	=	$r = 3.957$
2. Find v:	**1**	
	−	
	MR	
	=	
	←→	
	XM	
	+	
	1	
	÷	
	MR	
	=	
	f $\sqrt{x}$	
	MC	
	M+	
Enter E	**[68.971 1]**	
	÷	
	2	
	=	
	f tan	
	×	
	MR	
	=	
	f $\tan^{-1}$	
	×	
	2	
Note v	=	$v = +76°.988\ 62$. (If t is negative, v is also negative)
3. Find remaining terms for x_1 etc.:		
Enter Ω	**[77.102 4]**	
	MC	
	M+	
Note A	**f cos**	$\cos\Omega = 0.223\ 21 = A$
	MR	
Note B	**f sin**	$\sin\Omega = 0.974\ 771 = B$
Enter v	**[76.988 62]**	
	+	
Add ω	**[90.219 5]**	
	=	
	MC	
	M+	

Method 2A *continued*

Note C	**f cos**	$\cos(v + \omega) = -0.975\ 181 = C$
	MR	
Note D	**f sin**	$\sin(v + \omega) = 0.221\ 41 = D$
Enter i	[6.641 3]	
	MC	
	M+	
Note E	**f cos**	$\cos i = 0.993\ 29 = E$
	MR	
Note F	**f sin**	$\sin i = 0.115\ 653 = F$
4. Find x_1:		
Enter B	**[0.974 771]**	
	×	
Enter D	**[0.221 41]**	
	×	
Enter E	**[0.993 29]**	
	=	
	MC	
	M+	
Enter A	**[0.223 21]**	
	×	
Enter C	**[0.975 181]**	
	[CS]	C is negative
	−	
	MR	
	×	
Enter r	[3.957]	
Note x_1	=	$x_1 = -1.709\ 605\ 6$
5. Find y_1:		
Enter B	**[0.974 771]**	
	×	
Enter C	**[0.975 181]**	
	[CS]	C is negative
	=	
	MC	
	M+	
Enter A	**[0.223 21]**	
	×	
Enter D	**[0.221 41]**	
	×	
Enter E	**[0.993 29]**	
	+	
	MR	
	×	
Enter r	[3.957]	
	MC	
	M+	
Note y_1	=	$y_1 = -3.567\ 191\ 5$
6. Find z_1:		
Enter D	**[0.221 41]**	
	×	
Enter F	**[0.115 653]**	
	×	
	MR	
Note z_1	=	$z_1 = 0.101\ 325\ 7$

Method 2A *continued*

7. Note x ($= x_1$ from Step 4)	[no operation]	$x = -1.709\ 605\ 6$
8. Enter ϵ_{1950} in degrees	**23.445 788**	
	MC	
	M+	
Note G	**f cos**	$\cos\epsilon = 0.917\ 437 = G$
	←→	
	XM	
Note H	**f sin**	$\sin\epsilon = 0.397\ 881 = H$
9. Find y:		
Enter y_1 (Step 5)	**[3.567 191 5]**	
	[CS]	y_1 is negative
	×	
	MR	
	=	
	MC	
	M+	
Enter z_1 (Step 6)	**[0.101 325 7]**	
	×	
Enter H	**[0.397 881]**	
	=	
	←→	
	XM	
	−	
	MR	
Note y	=	$y = -3.312\ 988\ 9$
10. Find z:		
Enter y_1 (Step 5)	**[3.567 191 5]**	
	[CS]	y_1 is negative
	×	
Enter H	**[0.397 881]**	
	=	
	MC	
	M+	
Enter z_1 (Step 6)	**[0.101 325 7]**	
	×	
Enter G	**[0.917 437]**	
	+	
	MR	
Note z	=	$z = -1.326\ 357\ 8$
11. Find ξ:		
Enter x (Step 7)	**[1.709 605 6]**	
	[CS]	x is negative
	+	
Enter X_{1950}	**[0.437 278 5]**	
Note ξ	=	$\xi = -1.272\ 327\ 1$
12. Find η:		
Enter y (Step 9)	**[3.312 988 9]**	
	[CS]	y is negative
	+	
Enter Y_{1950}	**[0.808 554 9]**	
	[CS]	Y is negative
Note η	=	$\eta = -4.121\ 543\ 8$
13. Find ζ:		
Enter z	**[1.326 357 8]**	

Method 2A *continued*

	[CS]	z is negative
	+	
Enter Z_{1950}	[0.350 602 0]	
	[CS]	Z is negative
Note ζ	=	$\zeta = -1.676\,959\,8$
14. Find α:		
Enter η	[4.121 543 8]	
	[CS]	η is negative
	÷	
Enter ξ	[1.272 327 1]	
	[CS]	ξ is negative
	=	
	f $\tan^{-1}$	
	÷	
	15	
	=	
	STOP	(4.856 291 3)

If $\eta+$, $\xi+$, α is between 0^h and 6^h; if $\eta+$, $\xi-$, α is between 6^h and 12^h; if $\eta-$, $\xi-$, α is between 12^h and 18^h; if $\eta-$, $\xi+$, α is between 18^h and 24^h. If necessary, adjust display in multiples of 6^h until α appears in the correct quadrant.

Adjust α	[+ 12 =]	η is negative; ξ is negative. α is between 12^h and 18^h
	MC	
Note α in hours	M+	$\alpha = 16^h.856\,298$
Deduct integral hours; convert to minutes	−	
	[16]	
	×	
	60	
Note minutes	=	$51^m.378$
		$\alpha = 16^h\,51^m.38$
15. Find and store $\Delta\cos\delta$:		
Enter ξ	[1.272 327 1]	
	[CS]	ξ is negative
	←→	
	XM	
	×	
	15	
	=	
	f cos	
	←→	
	XM	
	÷	
	MR	
	=	
	MC	
	M+	
16. Find δ:		
Enter ζ	[1.6769598]	
	[CS]	ζ is negative
	÷	
	MR	
	=	

Method 2A *continued*

Note δ in degrees	**f tan⁻¹**	δ = –21°.244 72
Deduct integral degrees;	**–**	
convert to minutes	**[21]**	
	[CS]	CS if Southern dec.
	×	
	60	
Note minutes	**=**	14′.68
		δ = –21° 14′.7
17. Find Δ:		
Enter δ in degrees	**[21.244 72]**	
	[CS]	For Southern dec.
	f cos	
	f $\frac{1}{x}$	
	×	
	MR	
Note Δ	**=**	Δ = 4.627 9

Result 2A. The equatorial coordinates of Comet Smirnova-Chernykh 1975*e* at 0^h ET on 1977, January 17 were α = 16^h 51^m.38, δ = –21° 14′.7. The radius vector from the centre of the Sun, *r*, was 3.957 and the distance of the comet from the centre of the Earth, Δ, was 4.628 AU. As a check, the coordinates published in the 1977 *BAA Handbook* were α = 16^h 51^m.34, δ = –21° 14′.6, *r* = 3.958 and Δ = 4.628. The differences are slight: in α, +0^m.04; in δ, +0′.1; in *r*, –0.001; in Δ, nil.

Method 2B.

1. Find *r*:		
Enter *E* in degrees	**[68.971 1]**	
	STO 0	
	f cos	
Enter *e*	**[0.145 446]**	
	STO 1	
	×	
	CHS	
	1	
	+	
Enter *a*	**[4.174 405]**	
	×	
Note *r*	**STO 6**	*r* = 3.957
2. Find *v*:	**RCL 1**	
	1	
	+	
	1	
	RCL 1	
	–	
	÷	
	f √x	
	RCL 0	
	2	
	÷	
	f tan	

Method 2B *continued*

	×	
	g $\tan^{-1}$	
	2	
	×	
Check sign of v	STO 7	($v = +76°.988\ 689$). If t is negative, v is also negative
3. Terms of Eqn. 8.3		
Enter Ω	[77.102 4]	
	f cos	
	STO 0	
	f last x	
	f sin	
	STO 1	
Enter ω	[90.219 5]	
	RCL 7	
	+	
	f cos	
	STO 2	
	f last x	
	f sin	
	STO 3	
Enter i	[6.641 3]	
	f cos	
	STO 4	
	f last x	
	f sin	
	STO 5	
Enter ϵ_{1950}	23.445 788	
	STO 7	
4. Find α:	RCL 0	
	RCL 2	
	×	
	RCL 1	
	RCL 3	
	×	
	RCL 4	
	×	
	−	
	RCL 6	
	×	
	RCL 1	
	RCL 2	
	×	
	RCL 0	
	RCL 3	
	×	
	RCL 4	
	×	
	+	
	RCL 6	
	×	
	STO 1	
	$x \leftrightarrow y$	
	STO 0	

Method 2B *continued*

	RCL 6	
	RCL 3	
	RCL 5	
	×	
	×	
	STO 2	
	RCL 1	
	RCL 7	
	f cos	
	×	
	RCL 2	
	RCL 7	
	f sin	
	×	
	−	
	STO 3	
	RCL 1	
	RCL 7	
	f sin	
	×	
	RCL 2	
	RCL 7	
	f cos	
	×	
	+	
	STO 4	
	RCL 3	
	STO 1	
	RCL 4	
	STO 2	
Enter X_{1950}	[0.437 278 5]	
	STO + 0	
Enter Y_{1950}	[0.808 554 9]	
	[CHS]	Y is negative
	STO + 1	
Enter Z_{1950}	[0.350 602 0]	
	[CHS]	Z is negative
	STO + 2	
	RCL 1	
	RCL 0	
	g → P	
	R ↓	
	15	
	÷	
	STOP	(−7.143 693 293)
If display negative, add 24	[24 / +]	Adjustment made
	STO 7	
Note α	f H.MS	$\alpha = 16^h\ 51^m\ 22^s.93$
5. Find δ:	RCL 0	
	RCL 7	
	15	
	×	
	f cos	

Method 2B *continued*

	÷	
	RCL 2	
	$x \longleftrightarrow y$	
	STO 6	
	÷	
	g tan^{-1}	
	STO 5	
Note δ	**f H.MS**	$\delta = -21° \, 14' \, 41''.72$
6. Find Δ:	**RCL 5**	
	f cos	
	g $\frac{1}{x}$	
	RCL 6	
Note Δ	×	$\Delta = 4.627\,513$

Result 2B. The equatorial coordinates of Comet Smirnova-Chernykh 1975*e* at 0^h ET on 1977, January 17 were $\alpha = 16^h \, 51^m \, 22^s.93$ ($16^h \, 51^m.38$), $\delta = -21° \, 14' \, 41''.7$ ($-21° \, 14'.7$). The radius vector from the centre of the Sun, *r*, was 3.957, and the distance of the comet from the centre of the Earth, Δ, was 4.628 AU. As a check, the coordinates published in the 1977 *BAA Handbook* were $\alpha = 16^h \, 51^m.34$, $\delta = -21° \, 14'.6$, $r = 3.958$ and $\Delta = 4.628$. The differences are slight: in α, $+0^m.04$; in δ, $+0'.1$; in *r*, -0.001; in Δ, nil.

For further practice, try the following:

(a) From the following elements for Comet Arend-Rigaux 1950 VII compute α, δ, *r* and Δ for 0^h ET on 1977, December 13:

$T =$ 1978, February 2.416 4 ET
$\omega = 328°.986\,7$
$\Omega = 121°.524\,6$ } 1950.0
$i = 17°.855\,9$
$e = 0.599\,545$
$a = 3.600\,117$ AU
$n° = 0.144\,287\,6$
$P = 6.831$ years

The geocentric equatorial rectangular coordinates of the Sun for the required date (related to epoch 1950.0) are $X = -0.162\,825\,9$, $Y = -0.890\,781\,6$, $Z = -0.386\,248\,9$.

(b) *BAA Circular No. 570*, issued 1975, December 3, gave the following parabolic elements for the new Comet West 1975*n*:

$T =$ 1976, February 25.199 0 ET
$q = 0.196\,626$ AU
$\omega = 358°.419\,8$
$\Omega = 118°.226\,2$ } 1950.0
$i = 43.060\,1$

Compute α, δ, *r* and Δ for 0^h ET on 1976, July 1. Use the following coordinates of the Sun (1950.0), $X = -0.157\,601\,3$, $Y = 0.921\,525\,1$, $Z = 0.399\,585\,5$.

(c) The elements for the 1977/78 reappearance of Comet Tempel (1) 1867 II are:

$T = $ 1978, January 11.017 6 ET

$$\left.\begin{aligned} \omega &= 179^\circ.078\,3 \\ \Omega &= 68^\circ.339\,0 \\ i &= 10^\circ.544\,9 \end{aligned}\right\} 1950.0$$

$e = 0.519\,499$

$a = 3.115\,209$ AU

$n^\circ = 0.179\,255\,8$

$P = 5.498$ years

Compute α, δ, r and Δ for 0^h ET on 1977, July 16. The geocentric equatorial rectangular coordinates of the Sun for the required date: $X = -0.396\,701\,7$, $Y = 0.858\,556\,3$, $Z = 0.372\,279\,0$.

Your results should be:

(a) $\alpha = 2^h\,08^m\,40^s.74$ ($2^h\,08^m.68$), $\delta = -20^\circ\,33'\,17''.64$ ($-20^\circ\,33'.3$), $r = 1.548$ AU, $\Delta = 0.835$ AU. The ephemeris for Comet Arend-Rigaux 1950 VII in the 1977 *BAA Handbook* quotes the same values. If you did not obtain this result, check that you used $-51.416\,4$ for t and $-18^\circ.078\,885$ for E.

(b) At 0^h ET on 1976, July 1, the coordinates for Comet West 1975*n* were: $\alpha = 17^h\,52^m\,06^s.35$ ($17^h\,52^m.11$), $\delta = +11^\circ\,34'\,15''.44$ ($+11^\circ\,34'.3$), $r = 2.595$ AU, $\Delta = 1.706$ AU. *BAA Circular No. 570* quotes the same values. If you did not obtain this result, check that you used $+126.801\,0$ for t (1976 was a leap year) and $v = 148^\circ.042\,6$.

(c) The coordinates for Comet Tempel (1) 1867 II at 0^h ET on 1977, July 16 were: $\alpha = 10^h\,00^m\,41^s.09$ ($10^h\,00^m.68$), $\delta = +20^\circ\,28'\,24''.4$ ($20^\circ\,28'.4$), $r = 2.236$ AU, $\Delta = 3.019$ AU. The 1977 *BAA Handbook* gives the same values, with the exception of α, which is quoted as $10^h\,00^m.69$, a difference of $0^m.01$.

Note: the values for X, Y, Z (1950.0) used in these examples were obtained from the *AE*. See Chapter 9, Topics 3 and 4, for the method of computing approximations for X, Y, Z. In the Appendix there is a programme for the HP-67 which gives more accurate values (Programme 23).

NOTES

9 Approximations

Topic 1 Besselian Day Numbers.
Topic 2 Equation of the Equinoxes.
Topic 3 Geocentric rectangular equatorial coordinates, X, Y, Z, of the Sun.
Topic 4 Reduction of X, Y, Z for the mean equator and equinox of year (from Topic 3) to the mean equator and equinox for 1950.0.
Topic 5 Apparent geocentric equatorial coordinates α, δ for Mercury, Venus, Mars, Jupiter and Saturn.

Anyone who studies the *Explanatory Supplement to the AE* will soon realize that the values of certain items cannot be computed both quickly and accurately. Particular examples are the nutation in longitude ($\Delta\psi$) and the nutation in obliquity ($\Delta\epsilon$): in the former there are 69 terms, 46 of which are of short period, while in the latter there are 40 terms, all of which have a coefficient of $0''.000\,2$ or greater. Thus some sort of compromise is often desirable between the accuracy of the result and the speed of computation. It is hoped the topics in this Chapter will provide an acceptable compromise when circumstances are right, permitting reasonably accurate approximations to be achieved in the minimum of time.

1 **To compute approximations for the Besselian Day Numbers A, B, C, D, E, J and J', to an accuracy of $\pm 0''.05$ for A and B, $\pm 0''.75$ for C and D, and correct to 4 decimal places for E, J and J'.**

Introduction: The Besselian Day Numbers are used in the reduction of the mean place of a star from the start of a current Besselian solar year to the apparent place at any time during the year (see Chapter 4, Topic 1). The values for each day at 0^h ET are tabulated in the *AE.* There may be occasions when the Day Numbers are required in advance of publication of the ephemeris, or when it would be more appropriate to calculate the values for some other time than 0^h ET rather than to perform a series of interpolations. The examples will show how reasonably accurate values can be obtained.

In working the examples, no distinction is made between the two logic systems, algebraic and RPN, and so no keyboard entries are given. The reader must interpret the correct key-strokes required to suit his own calculator. Also, in order to avoid constant referring back, equations are not given at the start of the topic but are introduced as required in the working.

Further information: The *Explanatory Supplement to the AE* gives in-depth treatment, in particular Sections 2C and 5D; W. M. Smart, *Spherical Astronomy*, pp 242–247.

Example 1. Compute the Besselian Day Numbers *A, B, C, D, E, J and J′* for 1977, February 6 at 0^h ET. Assume the reduction is for a star with $\alpha = 2^h 00^m$.

Method

1. Compute τ.

Besselian solar years are counted from 1900, January $0^d.813$ ET in tropical years of 365.242 198 8 ephemeris days. To the start of the 1977 Besselian solar year there are 77 × 365 + 19 (leap days) = 28 124 days. Divide by 365.2422 = 77.00096 tropical years. The excess over 77 tropical years is 0.00096 year, which, multiplied bv 365, gives an excess of 0.350 days. Thus, the Besselian solar year 1977.0 commenced 0.350 days earlier than January $0^d.813$, i.e., January $0^d.463$, from which τ is reckoned.

τ = the number of days from the *nearest* beginning of a Besselian solar year ÷ 365.242 2. (*Note*—from July 1 to the end of the year the count starts backwards from the start of the following year, so τ is negative in that case.)

$$\tau = \frac{31 + 6 - 0.463}{365.242\,2} = +0.100\,0 \quad \text{(The } AE \text{ quotes the same value.)}$$

2. Compute $\Delta\psi$ (approximate).

$$\Delta\psi = -(17''.233 + 0''.017\,T)\sin\Omega + 0''.209\sin 2\Omega - 1''.273\sin 2L + 0''.126\sin g - 0''.204\sin 2☾ + 0''.068\sin l \qquad (9.1)$$

where $\Delta\psi$ = the nutation in longitude (containing 69 terms for the true value)
Ω = the longitude of the mean ascending node of the lunar orbit on the ecliptic
L = the geometric mean longitude of the Sun
g = the mean anomaly of the Earth ($L - \Gamma$)
☾ = the mean longitude of the Moon
l = the mean anomaly of the Moon (☾ $- \Gamma'$)
T = the interval elapsed since 12^h ET on 1900, January 0, expressed in Julian centuries of 36 525 ephemeris days
d = T expressed in days.

First, find d and T:

+77 × 365 + 19 (leap days) + 31 (January 1977) + 6 (February) − 0.5 = 28 160.5 (= d) ÷ 36 525 = 0.770 992 5 Julian centuries (= T).

In turn, solve for Ω, and take the sine and cosine
2Ω, and take the sine and cosine
L, and take the sine and cosine
$2L$, and take the sine and cosine
g, and take the sine
2☾, and take the sine and cosine
l, and take the sine.

In this topic we can use simplified terms of the mean orbital elements:

$$\Omega = 259^\circ.183\,3 - 0^\circ.052\,954\,d + 0^\circ.002\,078\,T^2 \qquad (9.2$$

$$L = 279^\circ.696\,7 + 0^\circ.985\,647\,d + 0^\circ.000\,303\,T^2 \tag{9.3}$$
$$g = 358^\circ.475\,8 + 0^\circ.985\,600\,d - 0^\circ.000\,150\,T^2 \tag{9.4}$$
$$2☾ = 180^\circ.868\,3 + 26^\circ.352\,793\,d - 0^\circ.002\,267\,T^2 \tag{9.5}$$
$$l = 296^\circ.104\,6 + 13^\circ.064\,992\,d + 0^\circ.009\,192\,T^2 \tag{9.6}$$

The angles computed may sometimes exceed 360°. Consult your calculator handbook first; some of the very cheap scientific-type calculators cannot perform functions on angles outside the range 0° to 90°, and extra care will have to be taken. Users of HP calculators and other advanced models will not experience any trouble in this respect. Where a record of the calculation does not need to be written out stage by stage there is no need for any manipulation of angles exceeding 360°. The sub-totals can be carried in the calculator memory and, although in excess of 360°, the correct sines and cosines will still be obtained.

To make this point clear, in Step 2 only of the computation the terms of the orbital elements are set out here in two columns below; on the left are the values as accumulated in the calculator memory, on the right are the same values as they would be set out in a written record of the computation, i.e., positive values of the total within the limits of 0° and 360°.

$\Omega =$	$259^\circ.183\,3$	$\Omega =$	$259^\circ.183\,3$
	$-\ 1\,491^\circ.211\,1$		$-\ 51^\circ.211\,1$
	$+\ 0^\circ.001\,2$		$+\ 0^\circ.001\,2$
$\Omega =$	$-\ 1\,232^\circ.026\,6$	$\Omega =$	$207^\circ.973\,4$
$\sin\Omega =$	$-0.469\,1$	$\sin\Omega =$	$-0.469\,1$
$\cos\Omega =$	$-0.883\,2$	$\cos\Omega =$	$-0.883\,2$
$\sin 2\Omega =$	$+0.828\,5$	$\sin 2\Omega =$	$+0.828\,5$
$\cos 2\Omega =$	$+0.560\,0$	$\cos 2\Omega =$	$+0.560\,0$
$L =$	$279^\circ.696\,7$	$L =$	$279^\circ.696\,7$
	$+\ 27\,756^\circ.312\,3$		$+\ 36^\circ.312\,3$
	$+\ 0^\circ.000\,2$		$+\ 0^\circ.000\,2$
$L =$	$+\ 28\,036^\circ.009\,2$	$L =$	$316^\circ.009\,2$
$\sin L =$	$-0.694\,5$	$\sin L =$	$-0.694\,5$
$\cos L =$	$+0.719\,5$	$\cos L =$	$+0.719\,5$
$\sin 2L =$	$-0.999\,4$	$\sin 2L =$	$-0.999\,4$
$\cos 2L =$	$+0.035\,2$	$\cos 2L =$	$+0.035\,2$
$g =$	$358^\circ.475\,8$	$g =$	$358^\circ.475\,8$
	$+\ 27\,754^\circ.988\,8$		$+\ 34^\circ.988\,8$
	$-\ 0^\circ.000\,1$		$-\ 0^\circ.000\,1$
$g =$	$+\ 28\,113^\circ.464\,5$		$393^\circ.464\,5$
			$-\ 360^\circ$
		$g =$	$33^\circ.464\,5$
$\sin g =$	$+0.551\,4$	$\sin g =$	$+0.551\,4$
2☾ $=$	$180^\circ.868\,3$	2☾ $=$	$180^\circ.868\,3$
	$+\ 742\,107^\circ.827\,3$		$+\ 147^\circ.827\,3$
	$-\ 0^\circ.001\,3$		$-\ 0^\circ.001\,3$
2☾ $=$	$+\ 742\,288^\circ.694\,3$	2☾ $=$	$328^\circ.694\,3$

$\sin 2☾ = -0.519\,6$ | $\sin 2☾ = -0.519\,6$
$\cos 2☾ = +0.854\,4$ | $\cos 2☾ = +0.854\,4$

$$
\begin{aligned}
l &= 296^\circ.104\,6 \\
&+ 367\,916^\circ.707\,2 \\
&+ 0^\circ.005\,5 \\
l &= +368\,212^\circ.817\,3
\end{aligned}
\qquad
\begin{aligned}
l &= 296^\circ.104\,6 \\
&+ 356^\circ.707\,2 \\
&+ 0^\circ.005\,5 \\
&\quad 652^\circ.817\,3 \\
&- 360^\circ \\
l &= 292^\circ.817\,3
\end{aligned}
$$

$\sin l = -0.921\,7$ | $\sin l = -0.921\,7$

It will be seen that much time can be saved by following the method on the left, allowing the terms to accumulate in the memory regardless of the number of revolutions; the desired end results, the sines and cosines, are identical. You will first have to test your calculator to confirm that this is possible. Users of the HP series will not need to make the test.

From Eqn. 9.1,

$$
\begin{aligned}
\Delta\psi = &-[(17''.233 + 0''.013) \times (-0.469\,1)] + (0''.209 \times 0.828\,5) \\
&-[1''.273 \times (-0.999\,4)] + (0''.126 \times 0.551\,4) \\
&-[0''.204 \times (-0.519\,6)] + [0''.068 \times (-0.921\,7)] \\
= &+9''.648.
\end{aligned}
$$

The 1977 *AE* (p 20) gives $+9''.696$; the error is thus $-0''.048$. The computation for $\Delta\psi$ has been greatly reduced, from 69 terms to 6, but the accuracy of the approximation is quite good.

3. Find B, from:

$$B = -\Delta\epsilon$$

where $\Delta\epsilon$ is the nutation in the obliquity, and

$$
\begin{aligned}
\Delta\epsilon &= (9''.210 + 0''.000\,9\,T)\cos\Omega - 0''.090\cos 2\Omega \\
&\quad + 0''.552\cos 2L + 0''.088\cos 2☾ \qquad (9.7) \\
&= [9''.211 \times (-0.883\,2)] - (0''.090 \times 0.560\,0) \\
&\quad + (0''.552 \times 0.035\,2) + (0''.088 \times 0.854\,4) \\
&= -8''.091\ (= -0^\circ.002\,247)
\end{aligned}
$$

$$B = +8''.091$$

4. Find ϵ, from:

$$\epsilon = 23^\circ.452\,294 - 0^\circ.013\,012\,5\,T \qquad (9.8)$$

where ϵ is the mean obliquity of the ecliptic, plus $\Delta\epsilon$ from Step 3, giving the mean obliquity of date.

$$
\begin{aligned}
&\quad 23^\circ.452\,3 \\
&- 0^\circ.010\,0 \\
&- 0^\circ.002\,2 \\
\epsilon = &\quad 23^\circ.440\,1
\end{aligned}
$$

$\sin\epsilon = +0.397\,8$ (for Step 5)
$\cos\epsilon = +0.917\,5$ (for Step 6)

5. Find A, from:

$$A = n\,\tau + \sin\epsilon\,\Delta\psi \qquad (9.9)$$

where n is the annual precession in dec. in seconds of arc, (from Chapter 2, Topic 3)

$= (20''.040\,3 \times 0.100\,0) + (0.397\,8 \times 9''.648)$

$= +\,5''.842.$

6. Approximate for C, from:

$$C = -k \cos\epsilon \cos L \tag{9.10}$$

where k is the constant of aberration ($20''.496$)

$= -20''.496 \times 0.917\,5 \times 0.719\,5$

$= -13''.530.$

7. Approximate for D, from:

$$D = -k \sin L \tag{9.11}$$

$= -20''.496 \times (-0.694\,5)$

$= +14''.234.$

8. Find E, from:

$$E = \lambda' \frac{\Delta\psi}{\psi'} \tag{9.12}$$

where λ', the planetary precession, $= 0''.124\,7 - 0''.018\,8\,T$

ψ', the luni-solar precession, $= 50''.370\,8 + 0''.005\,0\,T$

T, in both cases, being measured in tropical centuries from 1900.0.

$$E = 0''.110\,2 \times \frac{9''.648}{50''.374\,7}$$

$= +0''.021\,1$

$\div\ 15 = +0^s.001\,4.$

9. Find J and J' for the second-order corrections:

For Northern declinations:

$$\left.\begin{aligned} P_1 &= (A + D)\sin\alpha + (B + C)\cos\alpha \\ P_2 &= (A + D)\cos\alpha - (B + C)\sin\alpha \\ J &= 0.000\,005\,(P_1 P_2) \\ J' &= -0.000\,005\,\frac{(P_1)^2}{2} \end{aligned}\right\} \tag{9.13}$$

For Southern declinations:

$$\left.\begin{aligned} Q_1 &= (A - D)\sin\alpha + (B - C)\cos\alpha \\ Q_2 &= (A - D)\cos\alpha - (B - C)\sin\alpha \\ J &= 0.000\,005\,(Q_1 Q_2) \\ J' &= -0.000\,005\,\frac{(Q_1)^2}{2} \end{aligned}\right\} \tag{9.14}$$

It will be seen that J and J' vary according to the RA of the star. In this example, for a star of RA $= 2^h$, in Northern declination:

$P_1 = 10''.038\,0 + (-4''.710\,3)$

$= +\,5''.327\,7$

$P_2 = 17''.386\,3 - (-2''.719\,5)$

$= +20''.105\,8$

$J = +\ 0''.000\ 5 = +0^s.000\ 03$
$J' = -\ 0''.000\ 1.$

Result 1: The Besselian Day Numbers for 1977, February 6 at 0^h ET are:

	Computed (approximate)	*AE*	Error
A	$+\ 5''.842$	$+\ 5''.862$	$-0''.020$
B	$+\ 8''.091$	$+\ 8''.085$	$+0''.006$
C	$-13''.530$	$-13''.767$	$+0''.237$
D	$+14''.234$	$+13''.943$	$+0''.291$
E	$+0^s.001\ 4$	$+0^s.001\ 4$	Nil
J	$+0^s.000\ 03$	$+0^s.000\ 03$	Nil
J'	$-\ 0''.000\ 1$	$-\ 0''.000\ 1$	Nil

The approximate values computed for the Day Numbers are within the limits of accuracy set, and the objective of obtaining reasonably accurate values in a minimum of time is also achieved, without recourse to digital-computer facilities.

Note that although I have given the method for approximating values for all the Besselian Day Numbers, in the light of the magnitude of the errors likely to arise in *C* and *D*, it would be meaningless to attempt to employ these values for determining the apparent place of a star to the second order. Use of the approximate Day Numbers should therefore be restricted to reductions to the first order only; *J* and *J'*, although they can both be evaluated without error, can be ignored and the terms containing them in Eqns. 4.1 and 4.2 may be dropped.

Example 2.

In the previous example we saw how to calculate approximate values of the Besselian Day Numbers for 0^h ET at any date. There are, however, many instances when the Day Numbers are required, not for 0^h, but (for instance) at the time of upper transit of a particular star. In Topic 1 of Chapter 4, the Day Numbers for 1977, November 11.937 were found by interpolation between the values for 0^h on November 11 and 12. Now, if approximate values of the Day Numbers must be calculated for other than 0^h ET, they can be obtained directly without the need for interpolation.

For example, suppose we decide to compare approximate values of the Day Numbers for 1977, November 11.937 with the interpolated values actually employed in Chapter 4. There is no need to compute values for November 11 and 12 and then to interpolate to 11.937. The work can be considerably reduced by computing the values for November 11.937 directly, but still using the method of Example 1.

1. Compute τ.

As the date for which the information is required falls in the second half of the year, τ is counted backwards from the start of the next Besselian solar year, which is found by the method of Example 1 to commence on 1978, January $0^d.705$.

$$\tau = -\frac{31\ (\text{Dec}) + 19\ (\text{Nov}) - 0.937 + 0.705}{365.242\ 2}$$

$$= -0.136\ 3$$

2. Compute $\Delta\psi$ (approximate).

In this case, $d = 77 \times 365 + 19$ (leap days) $+ 315.937 - 0.5 = 28\,439.437$.

$$T = \frac{d}{36\,525}$$
$$= 0.778\,629\,4.$$

From Eqns. 9.2 to 9.6:

$$\begin{aligned}
\Omega &= 259^\circ.183\,3 \\
&\; - 1\,505^\circ.981\,9 \\
&\; + 0^\circ.001\,3 \\
&= -1\,246^\circ.797\,3 \quad \text{(i.e., } 193^\circ.202\,7\text{)} \\
\sin\Omega &= -0.228\,4 \\
\cos\Omega &= -0.973\,5 \\
\sin2\Omega &= +0.444\,7 \\
\cos2\Omega &= +0.895\,7
\end{aligned}$$

$$\begin{aligned}
L &= 279^\circ.696\,7 \\
&\; + 28\,031^\circ.245\,8 \\
&\; + 0^\circ.000\,2 \\
L &= 28\,310^\circ.942\,6 \quad \text{(i.e., } 230^\circ.942\,6\text{)} \\
\sin L &= -0.776\,5 \\
\cos L &= -0.630\,1 \\
\sin2L &= +0.978\,6 \\
\cos2L &= -0.206\,0
\end{aligned}$$

$$\begin{aligned}
g &= 358^\circ.475\,8 \\
&\; + 28\,029^\circ.909\,1 \\
&\; - 0^\circ.000\,1 \\
g &= +28\,388^\circ.384\,8 \quad \text{(i.e., } 308^\circ.384\,8\text{)} \\
\sin g &= -0.783\,9
\end{aligned}$$

$$\begin{aligned}
2☾ &= 180^\circ.868\,3 \\
&\; + 749\,458^\circ.596\,3 \\
&\; - 0^\circ.001\,4 \\
2☾ &= +749\,639^\circ.463\,2 \quad \text{(i.e., } 119^\circ.463\,2\text{)} \\
\sin2☾ &= +0.870\,7 \\
\cos2☾ &= -0.491\,9
\end{aligned}$$

$$\begin{aligned}
l &= 296^\circ.104\,6 \\
&\; + 371\,561^\circ.016\,9 \\
&\; + 0^\circ.005\,6 \\
l &= +371\,857^\circ.127\,1 \quad \text{(i.e., } 337^\circ.127\,1\text{)} \\
\sin l &= -0.388\,7
\end{aligned}$$

From Eqn. 9.1,

$$\begin{aligned}
\Delta\psi &= +3''.939\,0 + 0''.092\,9 - 1''.245\,8 + (-0''.098\,8) - 0''.177\,6 \\
&\; + (-0''.026\,4) \\
&= +2''.483
\end{aligned}$$

3. From Eqn. 9.7,

$$\Delta\epsilon = -8''.966\,6 - 0''.080\,6 + (-0''.113\,7) + (-0''.043\,3)$$
$$= -9''.204\ (= -0^\circ.002\,557)$$
$$B = -\Delta\epsilon$$
$$= +9''.204$$

4. From Eqn. 9.8,

$$\epsilon = 23^\circ.452\,3$$
$$- 0^\circ.010\,1$$
$$- 0^\circ.002\,6 \quad (\Delta\epsilon \text{ from Step 3})$$
$$\epsilon = 23^\circ.439\,6$$
$$\sin\epsilon = +\ 0.397\,8$$
$$\cos\epsilon = +\ 0.917\,5$$

5. From Eqn. 9.9,

$$A = [20''.040\,2 \times (-0.136\,3)] + (0.397\,8 \times 2''.483)$$
$$= -1''.744$$

6. From Eqn. 9.10,

$$C = -20''.496 \times 0.917\,5 \times (-0.630\,1)$$
$$= +11''.849$$

7. From Eqn. 9.11,

$$D = -20''.496 \times (-0.776\,5)$$
$$= +15''.915$$

8. From Eqn. 9.12,

$$E = 0''.110\,1 \times \frac{2''.483}{50''.374\,7}$$
$$= +0''.005\,4$$
$$\div\ 15 = +0^s.000\,4$$

9. From Eqn. 9.13, for a northern star of RA = 2^h:

$$P_1 = (14''.171 \times 0.50) + (21''.053 \times 0.866\,0) = +25''.317\,4$$
$$P_2 = (14''.171 \times 0.866\,0) - (21''.053 \times 0.50) = +\ 1''.745\,6$$
$$J = +0''.000\,22 = +0^s.000\,01$$
$$J' = -0''.001\,6$$

Result 2: The approximate values of the Besselian Day Numbers for 1977, Novembe[r] 11.937, computed directly, compared with the interpolated *AE* values used i[n] Chapter 4, Topic 1, with the resulting errors, are:

	Computed (approximate)	*AE*, interpolated	Error
τ	−0.136 3	−0.136 3	Nil
A	− 1″.744	− 1″.749	+0″.005
B	+ 9″.204	+ 9″.181	+0″.023
C	+11″.849	+12″.234	−0″.385
D	+15″.915	+15″.567	+0″.348

E	$+0^s.000\,4$	$+0^s.000\,4$	Nil
J	$+0^s.000\,01$	$+0^s.000\,01$	Nil
J'	$-0''.001\,6$	$-0''.001\,6$	Nil

The largest errors occur, as expected, in C and D, where we employ only coarse approximations. Anyone who feels that the errors in C and D are unacceptably large is recommended to consult pp 46–49 and 158–160 of the *Explanatory Supplement to the AE*. But for ordinary day-to-day working, to 0.1 second of time in RA and to the nearest second of arc in dec., the approximations for the Day Numbers are perfectly adequate. For example, if in the working of Method 1B of Chapter 4 the above approximate Day Numbers are employed in place of the interpolated *AE* values that we actually used, the equatorial coordinates α, δ for ϵ Cas at upper Greenwich transit on 1977, November 11.937 are found to be:

$$\alpha = 1^h\,52^m\,51^s.128 \qquad \delta = +63^\circ\,33'\,49''.14$$

When compared with the result obtained in Chapter 4, the errors introduced by using the approximations are in α $-0^s.023$, and in δ $+0''.36$.

The observer must decide for himself whether the time saved by calculating approximate values for the Day Numbers in the absence of the relevant ephemeris is justified or not in the light of the accuracy limits demanded by the type of job on which he is working. In any event he should regard any reductions to apparent place carried out by means of approximate Day Numbers as being to the first order only. (See the note at the end of Example 1.)

For further practice:

(a) Compute approximations for the Besselian Day Numbers A, B, C, D and E for 0^h ET on 1977, March 8.

(b) Given the interpolated Day Numbers $A = -3''.007$, $B = +8''.546$, $C = +17''.627$, $D = +7''.154$, $E = +0^s.00\,5$, $J = +0^s.000\,05$, $J' = -0''.001\,6$, $\tau = -0.215\,8$ for 1977, October 13.9, and the upper Greenwich transit of γ Cep (RA$=$ $23^h\,38^m$), compute approximations for the Day Numbers and tabulate the errors.

Your answers should be:

(a) $A = +6''.988$, $B = +7''754$, $C = -18''.213$, $D = +5''.106$, $E = +0^s.001\,2$. As a check, the *AE* gives $A = +6''.982$, $B = +7''.741$, $C = -18''.339$, $D = +4''.497$, $E = +0^s.001\,2$.

(b) The approximate Day Numbers, with the errors (in the sense 'computed minus *AE*') in brackets, are:

$A = -3''.040$ ($-0''.033$)
$B = +8''.569$ ($+0''.023$)
$C = +17''.397$ ($-0''.230$)
$D = +7''.784$ ($+0''.630$)
$E = +0^s.000\,5$ (Nil)
$J = +0^s.000\,06$ ($+0^s.000\,01$)
$J' = -0''.001\,6$ (Nil)
$\tau = -0.215\,8$ (Nil)

2 To calculate an approximate value for the equation of the equinoxes for 0^h UT at any date, correct to within $\pm 0^s.005$.

Introduction: Reference to the correction for the equation of the equinoxes was made in Chapter 1, Topic 1, being the addition required to convert Greenwich Mean Sidereal Time into Apparent GST. Full evaluation of the correction entails lengthy computation because one of the factors involved, the nutation in longitude ($\Delta\psi$), contains 69 terms. Consequently, unless the relevant *AE* is held (where the value is listed daily) one has little inclination to compute the true value of the equation of the equinoxes. However, when it is essential to know the value in advance of publication of the *AE* it is possible quickly to compute an approximate value for any date. The method is demonstrated in this Topic.

The Equation:

$$E_E = \Delta\psi \cos\epsilon \qquad (9.15)$$

where $\Delta\psi$ = the nutation in longitude
ϵ = the obliquity of the ecliptic of date (the true obliquity).

Further information: *Explanatory Supplement to the AE*, Section 3C; D. McNally, *Positional Astronomy*, Chap. 5.1.

Example: Calculate the value of the equation of the equinoxes for 1978, January 0 at 0^h UT.

Method: The necessary computation has already been outlined in Steps 2, 3 and 4 of the examples in Topic 1 of this Chapter.

1. Find d and T. Then, from Eqns. 9.2 to 9.6, find the values for Ω, 2Ω, $2L$, g, $2☾$ and l, with their sines.

$d =$ 28 488.5 $\qquad T = 0.779\,972\,6$

$$\begin{aligned}
\Omega &= 259°.183\,3 \\
&\ - 1\,508°.580\,0 \\
&\ + 0°.001\,3 \\
\Omega &= - 1\,249°.395\,4 \quad \text{(i.e., } 190°.604\,4) \\
\sin\Omega &= -0.184\,0 \\
\sin 2\Omega &= +0.361\,8
\end{aligned}$$

$$\begin{aligned}
L &= 279°.696\,7 \\
&\ + 28\,079°.604\,6 \\
&\ + 0°.000\,2 \\
L &= 28\,359°.301\,5 \quad \text{(i.e., } 279°.301\,5) \\
\sin 2L &= -0.319\,0
\end{aligned}$$

$$\begin{aligned}
g &= 358°.475\,8 \\
&\ + 28\,078°.265\,6 \\
&\ - 0°.000\,1 \\
g &= 28\,436°.741\,3 \quad \text{(i.e., } 356°.741\,3) \\
\sin g &= -0.056\,8
\end{aligned}$$

$$2☾ = 180°.868\,3 + 750\,751°.543\,4 - 0°.001\,4$$

$$2☾ = 750\,932°.410\,3 \quad \text{(i.e., } 332°.410\,3\text{)}$$

$$\sin 2☾ = -0.463\,1$$

$$l = 296°.104\,6 + 372\,202°.024\,6 + 0°.005\,6$$

$$l = 372\,498°.134\,8 \quad \text{(i.e., } 258°.134\,8\text{)}$$

$$\sin l = -0.978\,6$$

2. From Eqn. 9.1:

$$\Delta\psi = [-(17''.233 + 0''.013) \times (-0.184\,0)] + 0''.209\,(0.361\,8) - 1''.273\,(-0.319\,0) + 0''.126\,(-0.056\,8) - 0''.204\,(-0.463\,1) + 0''.068\,(-0.978\,6)$$

$$= +3''.676$$

3. From Eqn. 9.8:

$$\epsilon = 23°.452\,3 - 0°.010\,1$$

$$= 23°.442\,2$$

(*Note:* there is no need here to evaluate $\Delta\epsilon$, because $\cos\epsilon$ to 4 decimal places would not be affected)

$$\cos\epsilon = +\,0.917\,5$$

4. From Eqn. 9.15:

$$E_E = +3''.676 \times 0.917\,5$$

$$= +3''.373$$

$$\div\, 15 = +0^s.225$$

Result: The approximate equation of the equinoxes for 1978, January 0 at 0^h UT is $+0^s.225$. The *AE* gives the value $+0^s.226$. The error in this case is $-0^s.001$, and the result is well within the accuracy limit set.

For further practice, compute the equation of the equinoxes for 0^h UT on

(a) 1977, March 8;
(b) 1977, February 6;
(c) 1977, July 3;
(d) 1977, September 6.

Your answers should be:

(a) $+0^s.513$. (The *AE* gives $+0^s.512$)
(b) $+0^s.598$. (The *AE* gives $+0^s.593$)
(c) $+0^s.416$. (The *AE* gives $+0^s.415$)
(d) $+0^s.338$. (The *AE* gives $+0^s.343$).

3 To compute (a) to three decimal places, the approximate geocentric equatorial rectangular coordinates X, Y, Z of the Sun, for the mean equator and equinox of year, at any time on any date, and (b) the approximate equatorial coordinates α, δ of the Sun, the radius vector, horizontal parallax and semi-diameter at the same time as (a).

Introduction: An outline of the difficulties to be overcome for high-precision work is given in Chapter 6. Where close approximations will suffice, the basic approach to the problem is to furnish reasonably accurate elements for the orbit of the Earth, and then to apply, with slight amendments, the method used in Chapter 8 for comets. That is the method demonstrated here.

When x, y, z have been evaluated the signs are changed and the values become those for X, Y, Z, related to the mean equator and equinox of year. If these are all that is required the calculation stops at this stage.

If further data are needed for the same date, such as the apparent geocentric coordinates α, δ of the Sun, the radius vector, horizontal parallax or semi-diameter, the computation continues as indicated.

The equations: Those which are employed have been given in Chapters 7 and 8.

Example 3. Data are required for the Sun at 12^h ET on 1978, March 16 (1978.206 7). Find X, Y, Z, apparent α, δ, R, HP and SD.

The elements of the Earth's orbit are:

$T = 0.008\,8$ of year (e.g., 1978.008 8)*
$\omega = 101°.220\,8 + 1°.719\,2\,T$
$\Omega = 0°$
$i = 0°$
$e = 0.016\,751 - 0.000\,04\,T$
$\epsilon = 23°.452\,29 - 0°.013\,0\,T$
$a = 1.0$ AU
$n° = 0.985\,61$
$q = 0.9833\,26$ AU
$P = 365.25\,d$

where T is the interval in centuries from 1900, January 0, with no distinction necessary between the Julian and tropical centuries.

First, evaluate ω, e and $e°$, finding $\omega = 102°.561\,8$, $e = 0.016\,72$ and $e° = 0.958\,0$. Then find the mean anomaly for the required time from $M = n°t$ and iterate for E in Eqn. 8.7. $t = 365.26\,(1978.206\,7 - 1978.008\,8) = 72.28$. Thus find $M = 71°.24$ and $E = 72°.15$. Because the orbit is almost circular it will always be found, when iterating for E, that E is close to M in value.

* This is not strictly true, as perihelion date alters slightly from year to year owing to a combination of the length of the anomalistic year ($365^d.259\,6$) and perturbations from the gravitational forces exerted by other planets, but it will be found in practice that the value shown can be employed for approximation work with little error.

Method 3A.

. Find R: Enter E	**[72.15]**	
	f cos	
	$\times$	
Enter e	**[0.016 72]**	
	M+	
	CS	
	+	
	1	
Note R	=	$R = 0.994\ 875$
. Find $(\nu + \omega)$ and store	**1**	
	−	
	MR	
	=	
	←→	
	XM	
	+	
	1	
	÷	
	MR	
	=	
	f $\sqrt{x}$	
	MC	
	M+	
Enter E	**[72.15]**	
	÷	
	2	
	=	
	f tan	
	$\times$	
	MR	
	=	
	f $\tan^{-1}$	
	$\times$	
	2	
	+	
Enter ω	**[102.561 8]**	
	=	$(\nu + \omega = 175°.626\ 02)$
	MC	
	M+	
Find X:	**f cos**	
	$\times$	
Enter R	**[0.994 875]**	
	CS	
Note X	=	$X = +0.991\ 977\ 9$ (+0.992, use of more than 3 decimal places is not justified) See Topic 4 for conversion to 1950.0
Find $R \sin(\nu + \omega)$ and ore	**MR**	
	f sin	
	$\times$	
Enter R	**[0.994 875]**	
	CS	

Method 3A *continued*

	=	
	MC	
	M+	
5. Find Y:		
Enter ϵ	**[23.442 2]**	
	f cos	
	×	
	MR	
Note Y	=	$Y = -0.069\,612\,5$ (-0.070)
6. Find Z:		
Enter ϵ	**[23.442 2]**	
	f sin	
	×	
	MR	
Note Z	=	$Z = -0.030\,184\,9$ (-0.030)
7. For further data for date:		
Enter Y in full	**[0.069 612 5]**	
	CS	
	÷	
Enter X in full	**[0.991 977 9]**	
	=	
	f tan^{-1}	
	÷	
	15	
	=	
	STOP	($-0.267\,611\,4$)

If $Y+$, $X+$, $0 < \alpha < 6^h$; if $Y+$, $X-$, $6 < \alpha < 12^h$; if $Y-$, $X-$, $12 < \alpha < 18$ if $Y-$, $X+$, $18 < \alpha < 24^h$.

If necessary, adjust display by multiples of 6^h until α appears in the corre quadrant.

8. Adjust	**[+ 24]**	Adjustment made
Note hours; deduct integral hours	=	$23^h.732389$
	−	
	[23]	
	×	
	60	
Note minutes: deduct integral minutes	=	43^m
	−	
	[43]	
	×	
	60	
Note seconds	=	57^s
		$\alpha = 23^h\ 43^m\ 57^s$
9. Find δ		
Enter α in hours	**[23.732 389]**	
	×	
	15	
	=	
	f cos	
	MC	
	M+	

Method 3A *continued*

Enter X in full	**[0.991 977 9]**	
	÷	
	MR	
	=	
	←→	
	XM	
Enter Z in full	**[0.030 184 9]**	
	CS	Z is negative
	÷	
	MR	
	=	
Note integral degrees; if display is negative,	**f tan^{-1}**	–1°
change sign	[CS]	
Deduct integral degrees	–	
	[1]	
	×	
	60	
Note minutes; deduct minutes	=	44′
	–	
	[44]	
	×	
	60	
Note seconds	=	19″
		δ = –1° 44′ 19″
10. Find HP:	**8.794**	
	÷	
Enter R (Step 1)	**[0.994 875]**	
	MC	
	M+	
Note HP	=	HP = 8″.84
11. Find SD:	**961.18**	
	÷	
	MR	
Note SD	=	SD = 966″.13
		= 16′ 06″.13

Result 3A. (a) The approximate rectangular coordinates for the Sun at 12^{h} ET on 1978, March 16 are X = +0.992, Y = –0.070, Z = –0.030.
(b) The radius vector at that time is 0.994 9 AU, the horizontal parallax is 8″.84 and the semi-diameter of the Sun is 16′ 06″.13.

For comparison with the *AE* values refer to the note with Result 3B which follows.

Method 3B.

1. Enter E	**[72.15]**	
	STO 0	
	f cos	
Enter e	**[0.016 72]**	
	STO 1	
	×	
	CHS	

Method 3B *continued*

	1	
	+	
	STO 6	($R = 0.994\,875$)
	RCL 1	
	1	
	+	
	1	
	RCL 1	
	−	
	÷	
	f $\sqrt{x}$	
	RCL 0	
	2	
	÷	
	f tan	
	×	
	g $\tan^{-1}$	
	2	
	×	($v = 73°.064\,3$)
2. Enter ω	**[102.561 8]**	
	+	
	f cos	
	STO 2	
	f last x	
	f sin	
	STO 3	
3. Enter ϵ	**[23.442 2]**	
	STO 7	
	RCL 2	
	RCL 6	
	×	
	CHS	
	STO 0	
	RCL 3	
	RCL 6	
	×	
	STO 1	
	RCL 7	
	f cos	
	×	
	STO 3	
	RCL 1	
	RCL 7	
	f sin	
	×	
	CHS	
	STO 2	
	RCL 3	
	CHS	
	STO 1	

Method 3B *continued*

X is now in R_0, Y in R_1, Z in R_2.

For X, RCL 0;	$X = +0.992$	(Use of more than 3 decimal places is not
Y, RCL 1;	$Y = -0.070$	justified. See Topic 4 for conversion to 1950.0.)
Z, RCL 2;	$Z = -0.030$	

---

4. For further data for date:	**RCL 1**	
	RCL 0	
	g → P	
	R ↓	
	15	
	÷	
	STOP	
5. If display negative, add 24	[**24** +]	Adjustment made
	STO 7	
Note α	**f H.MS**	$\alpha = 23^h\,43^m\,57^s$
Find δ:	**RCL 0**	
	RCL 7	
	15	
	×	
	f cos	
	÷	
	RCL 2	
	$x \longleftrightarrow y$	
	÷	
	g tan⁻¹	
Note δ	**f H.MS**	$\delta = -1°\,44'\,19''$
Note R	**RCL 6**	$R = 0.994\,875$
Find HP:	**8.794**	
	$x \longleftrightarrow y$	
Note HP	÷	HP $= 8''.84$
Find SD:	**961.18**	
	RCL 6	
Note SD	÷	SD $= 966''.13$
		$= 16'\,06''.13$

Result 3B. The approximate rectangular coordinates for the Sun at 12^h ET on 1978, March 16 are $X = +0.992$, $Y = -0.070$, $Z = -0.030$. The *AE* lists, for

March 16, $X = +0.991\,0$, $Y = -0.078\,8$, $Z = -0.034\,2$ (at 0^h)
March 17, $X = +0.992\,6$, $Y = -0.063\,0$, $Z = -0.027\,3$ (at 0^h)

The computed terms fall within the tabulated values, but interpolation between the *AE* values for 12^h would reveal slight errors.

The other data derived for the Sun at the same time compare as follows:

Computed, 12^h March 16	*AE*, 0^h March 16	*AE*, 0^h March 17
$\alpha = 23^h\,43^m\,57^s$	$23^h\,41^m\,49^s$	$23^h\,45^m\,28^s$
$\delta = -1°\,44'\,19''$	$-1°\,58'\,08''$	$-1°\,34'\,26''$
$R = 0.994\,875$	$0.994\,747$	$0.995\,015$
HP $= 8''.84$	$8''.85$	$8''.84$
SD $= 16'\,06''.13$	$16'\,06''.26$	$16'\,06''.00$

Again, the computed values fall between those tabulated in the *AE*, but interpolation would reveal small errors of about 20^s in RA and $2'$ in dec. Although the computed values relate to the mean equator and equinox of year, and the *AE* values are for the true equator and equinox, this factor does not account for the greater part of the errors.

The two methods 3A and 3B give identical results.

For further practice, try the following:

(a) Compute the approximate rectangular equatorial coordinates of the Sun for 0^h ET 1974, July 23 (1974.558 5).

(b) Compute the approximate equatorial coordinates α, δ, the radius vector, horizontal parallax and semi-diameter of the Sun for 0^h ET on 1977, December 22. Use the 1978 perihelion date. (t, M and E will be negative.)

(c) Compute the approximate equatorial coordinates α, δ, of the Sun for 1976, August 3 at 0^h ET (1976.591 4).

Your results should be:

(a)	$X = -0.505$	The *AE* gives	$X = -0.504\,7$
	$Y = +0.809$		$Y = +0.809\,0$
	$Z = +0.351$		$Z = +0.350\,8$
(b)	$\alpha = 18^h\,00^m\,36^s$	The *AE* gives	$\alpha = 18^h\,00^m\,07^s$
	$\delta = -23°\,26'\,32''$		$\delta = -23°\,26'\,22''$
	$R = 0.9837$		$R = 0.98367$
	$HP = 8''.94$		$HP = 8''.95$
	$SD = 16'\,17''.15$		$SD = 16'\,17''.14$
(c)	$\alpha = 8^h\,55^m\,10^s$	The *AE* gives	$\alpha = 8^h\,52^m\,57^s$
	$\delta = +17°\,22'\,52''$		$\delta = +17°\,31'\,47''$

The error in (c) is not due to the fact that 1976 was a leap year, as $t = 212.80$ includes the leap day. It is, in fact, due to the perihelion date being January 4.5. Our result is nearer the *AE* value for August 4, $\alpha = 8^h\,56^m\,48^s$, $\delta = +17°\,15'\,59''$. An error of this magnitude will be rare.

4 To convert the approximate equatorial rectangular coordinates of the Sun, *X*, *Y*, *Z*, from the equator and equinox of year to the equator and equinox of 1950.0, for use in the topics of Chapter 8.

Introduction: In the preceding topic a method was demonstrated of obtaining approximate values for X, Y, Z to 3 decimal places. These values relate to the equator and equinox of year (of calculation). In the comet computations of Chapter 8, where the elements are referred to the equator and equinox of 1950.0, the values of X, Y, Z must also be referred to that same epoch. An approximate transformation, no more rigourous than the method used to derive X, Y, Z in Topic 3, is given here. In Appendix II, an HP-67 programme is included (Programme 23), which gives highly accurate results.

The equations:

$$\left.\begin{aligned} X_{1950} &= X + 0.007\,8\ Y + 0.003\,4\,Z \\ Y_{1950} &= Y\ - 0.007\,8\ X \\ Z_{1950} &= Z\ - 0.003\,4\ X \end{aligned}\right\} \qquad (9.16)$$

The values of the coefficients are correct for 1985.0. They may be used for approximations over the period 1970–2000, without significant error for our purposes (mainly for the calculations of Chapter 8).

Further information: *Explanatory Supplement to the AE*, p 34 and Table 2.2.

Example 4. The equatorial rectangular coordinates of the Sun at 0^h ET on 1978, July 3 are, to 3 decimal places: $X = -0.189$, $Y = +0.916$, $Z = +0.397$. These relate to the true equator and equinox. Convert to relate to the equator and equinox of epoch 1950.0.

The workings are so simple they do not need to be shown here. Using Eqn. 9.16, we find:

$X_{1950} = -0.181$, $Y_{1950} = +0.917$, $Z_{1950} = +0.398$.

To 3 decimal places, the *AE* gives: $X_{1950} = -0.182$, $Y_{1950} = +0.918$, $Z_{1950} = +0.398$.

Try the following:

Convert the approximate result for further-practice problem (a) of the preceding topic to coordinates referred to 1950.0.

Your results should be: $X_{1950} = -0.497$, $Y_{1950} = +0.812$, $Z_{1950} = +0.352$.

5 To compute the approximate apparent geocentric equatorial coordinates α, δ, of Mercury, Venus, Mars, Jupiter and Saturn, at any desired ET on any date, plus the horizontal parallax, semi-diameter in seconds of arc, and distance from the Earth in astronomical units.

Introduction: See the introduction to Topic 3, and refer to Chapter 6 with regard to the difficulties encountered in high-precision work. The accuracy attained here is more than sufficient for finding purposes.

The orbit elements are given for epoch 1950.0, with annual variations in brackets. T is given in terms of the perihelion passage nearest to 1975.0. The results give good approximations over the period of 1950–2000, even for Mercury.

For greater accuracy, and over a range of about 3 000 years, see Programmes 25 to 27 and 38 in Appendix II. These programmes are for the HP-67 and HP-97.

The equations: The method employed is based on the equations of Chapter 8.

Further information: *Explanatory Supplement to the AE*, Section 4D.

Example 5(a): Mercury.

$T = 1975.078\,4 \pm$ multiples of P

$P = 0.240\,847$ tropical year

$$\left.\begin{aligned} \omega &= 28^\circ.938\,89\ (+0^\circ.003\,7) \\ \Omega &= 47^\circ.738\,55\ (+0^\circ.011\,9) \\ i &= 7^\circ.003\,81\ (+0^\circ.000\,02) \end{aligned}\right\}\ 1950.0$$

$e = 0.205\,624\ (+0.000\,000\,2)$, 1950.0

$e^\circ = 11^\circ.782$
$a = 0.387\,099$ AU
$n^\circ = 4^\circ.092\,339$ (per day)

$$\text{HP} = \frac{8''.794}{\Delta}$$

$$\text{SD} = \frac{3''.34}{\Delta}$$

Compute data for 1978, November 23, at 0^h ET (1978.895 3). In this case, the orbital period being about ¼ year, the nearest time of perihelion passage will be 15 revolutions of the planet later than T as tabulated, i.e., $T = 1975.078\,4 + 15P = 1978.691\,1$.

Update, where necessary, the elements to 1979.0 (the nearest epoch) and find, in the order that they will be required for the calculation:

$e = 0.205\,63$
$\Omega = 48^\circ.083\,65$
$\omega = 29^\circ.046\,19$
$i = 7^\circ.004\,39$
$\epsilon = 23^\circ.442\,0$

Also, by the method of Topic 3 of this Chapter, find approximations for X, Y, Z for the Sun at the required date, November 23 at 0^h ET. (*Note:* we have updated the elements to a current epoch, so we require X, Y, Z of date, *not* referred to epoch 1950.0, so no conversion is necessary.)

$X = -0.487 \qquad Y = -0.788 \qquad Z = -0.342$

We shall be using the same date for the other planets in the latter part of this topic, so once X, Y, Z have been computed for this example the same values can be used in the other examples.

To find the mean anomaly by Eqn. 8.7 we must first find t in days, which is 365.24 (1978.895 3) – 1978.691 1) = 74.58 days. Then $M = 74.58\,n^\circ = 305^\circ.21$, and, by substitution in Eqn. 8.7, $E = 294^\circ.49$.

The preparatory work has now been completed. For the purposes of the computation we treat Mercury in exactly the same way as a periodic comet.

Process the foregoing data by Method 2A or 2B of Chapter 8. The only minor change required is to ignore the subscript 1950 when entering ϵ and X, Y, Z.

We find that the position for Mercury at 0^h ET on 1978, November 23 is

$\alpha = 17^h\,20^m\,31^s \qquad \delta = -25^\circ\,16'\,10'' \qquad \Delta = 0.866\,7$ AU

Store the value for Δ in the calculator memory. Then, to find the horizontal parallax, divide $8''.794$ by Δ, and obtain HP = $10''.15$. To find the semi-diameter of Mercury at this date, divide $3''.34$ by Δ, and obtain SD = $3''.85$.

Now for the moment of truth! The values listed in the 1978 *AE* are: $\alpha = 17^h\,20^m\,11^s$, $\delta = -25^\circ\,15'\,11''$, $\Delta = 0.868\,031\,3$ AU, HP = $10''.14$, SD = $3''.85$. You must judge for yourself whether the computed approximations are close enough to the values tabulated in the *AE* to suit your purpose. You might, for example, be concerned about the longer-term usefulness of the method. Then let us use it to evaluate data for Mercury for 1950, July 25 at 0^h ET (1950.564 0). In this case the nearest perihelion passage is 102 revolutions earlier, so T becomes 1975.078 4 –

102 P = 1950.512 0. t = 365.24 (1950.564 0 – 1950.512 0) = 18.99 days. Thus, M = 77°.71, and E is found to be 89°.49. The orbit elements do not need updating—they already refer to 1950.0. All we need now, from Topic 3, are values for X, Y, Z. These are found to be X = –0.528, Y = +0.796, Z = +0.345. The computation for position is performed as before, and the result is found to be $\alpha = 9^h\,16^m.1$; δ = +17° 34′.0. The 1950 *BAA Handbook* gives: $\alpha = 9^h\,16^m.4$; δ = +17° 34′.

The method thus works for over 100 revolutions into the past and indicates that for approximate results (which are all that we set out to achieve) reasonably close values will be obtained during the period 1950 to 2000.

Example 5(b): Venus.

T = 1975.308 4 ± multiples of P
P = 0.615 21 tropical year
ω = 54°.637 92 (+0°.005 1) ⎫
Ω = 76°.229 65 (+0°.009 0) ⎬ 1950.0
i = 3°.394 13 (+0°.000 01) ⎭
e = 0.006 797 (–0.000 000 2), 1950.0
$e°$ = 0°.389 4
a = 0.723 332 AU
$n°$ = 1°.602 130 (per day)

$$\text{HP} = \frac{8''.794}{\Delta}$$

$$\text{SD} = \frac{8''.41}{\Delta}$$

Compute data for 1978, November 23, at 0^h ET (1978.895 3). The orbital period is about $\frac{2}{3}$ year, so the nearest time of perihelion passage will be 6 revolutions of the planet later than T as tabulated, *i.e.*, T = 1975.308 4 + 6P = 1978.999 7. t = 365.24 (1978.895 3 – 1978.999 7) = –38.12 days. Find M = –61°.07 and E = –61°.41. Now update, where necessary, the elements to the nearest epoch, 1979.0, and find, in the order they will be required for the calculation: e = 0.006 791, Ω = 76°.490 65, ω = 54°.785 82, i = 3°.394 42. From the Mercury example, we take ϵ = 23°.442 0, X = –0.487, Y = –0.788, Z = –0.342.

The preparatory work now complete, process the data by Method 2A or 2B of Chapter 8, as in the case of example 5(a) for Mercury, again ignoring subscripts 1950 when entering ϵ and X, Y, Z.

We find that the position for Venus at 0^h ET on 1978, November 23 is $\alpha = 14^h\,21^m\,07^s$, δ = –15° 10′ 15″, Δ = 0.2982 AU. Store Δ. To find the horizontal parallax, divide 8″.794 by Δ, and for the semi-diameter of the planet, divide 8″.41 by Δ, finding: HP = 29″.49, SD = 28″.20. For comparison, the values tabulated in the *AE* are: $\alpha = 14^h\,21^m\,03^s$, δ = –15° 07′ 44″, Δ = 0.297 970 9 AU, HP = 29″.53, SD = 28″.22. Refer to the comment on accuracy with the results for Mercury.

Example 5(c): Mars.

T = 1975.449 0 ± multiples of P
P = 1.880 89 tropical years

$$\left.\begin{aligned} \omega &= 285^\circ.966\,66\,(+0^\circ.010\,7) \\ \Omega &= 49^\circ.171\,92\,(+0^\circ.007\,7) \\ i &= 1^\circ.850\,00\,(\pm 0) \end{aligned}\right\} 1950.0$$

$$e = 0.093\,359\,(+0.000\,000\,9),\ 1950.0$$

$$e^\circ = 5^\circ.349\,1$$

$$a = 1.523\,69 \text{ AU}$$

$$n^\circ = 0^\circ.524\,033 \text{ (per day)}$$

$$\text{HP} = \frac{8''.794}{\Delta}$$

$$\text{SD} = \frac{4''.68}{\Delta}$$

Compute data for 1978, November 23, at 0^{h} ET (1978.895 3). Follow exactly th same procedure used in the previous examples for Mercury and Venus. There is nc perihelion passage during 1978; the nearest is 1979.209 4. Update ω, Ω and e tc 1979.0; then find t, M and E. Process as before, using the same values for ϵ, X, Y and Z. We find that $\alpha = 16^{\text{h}}\,56^{\text{m}}\,22^{\text{s}}$, $\delta = -23^\circ\,18'\,46''$, $\Delta = 2.395\,6$ AU, HP = $3''.67$, SD = $1''.95$. (Check, if required, $t = -115.23$, $M = -60^\circ.38$, $E = -65^\circ.24$.

The values tabulated in the 1978 *AE* are: $\alpha = 16^{\text{h}}\,55^{\text{m}}\,38^{\text{s}}$, $\delta = -23^\circ\,16'\,54''$ $\Delta = 2.396\,186\,9$ AU, HP = $3''.67$, SD = $1''.95$.

The Outer Planets.

Readers are advised to consult the *Explanatory Supplement to the AE* regardin the difference between mean elements and osculating elements, and the use of th latter in the preparation of ephemerides. So far in this topic we have employe mean elements and have seen that useful approximations can be derived for th geocentric positions of the inner planets.

The four giant planets whose orbits lie outside the asteroid belt account for mos of the planetary mass of the Solar System. At the distances involved, the gravi tational effect of the Sun is correspondingly weaker, and the mutual effects that th massive outer planets exert upon each other are greater than the perturbations the cause to the orbits of the smaller inner planets. The use of mean orbital elements i really no longer practicable in these circumstances.

Where a rough indication of position would be sufficient, that is to say, if error of up to half a degree are not of great consequence, I give mean elements fc Jupiter and Saturn so that the same method of computation may be employed, bu advise caution in their use. It will be seen, if the reader follows through from th preceding examples, that computation of the position for Jupiter on the date w have been using (1978, November 23) gives, fortuitously, a result very close to th *AE* apparent position; it will not always be so. For Saturn, the computation gives position which the planet does not reach until some 18 days later, although remains within the limit of accuracy desirable for finding purposes. As the resul for Uranus, Neptune and Pluto would be even more unreliable, no elements ar given for these planets.

Example 5(d): Jupiter.

$$T = 1975.613\,3 \pm \text{multiples of } P$$

$P = 11.862\ 23$ tropical years

$\left.\begin{aligned} \omega &= 273^\circ.573\ 75\ (+0^\circ.006\ 0) \\ \Omega &= 99^\circ.943\ 33\ (+0^\circ.010\ 1) \\ i &= 1^\circ.305\ 92\ (-0^\circ.000\ 06) \end{aligned}\right\}$ 1950.0

$e = 0.048\ 419\ 0\ (+0.000\ 001\ 6)$, 1950.0

$e^\circ = 2^\circ.774\ 2$

$a = 5.202\ 803$ AU

$n^\circ = 0^\circ.083\ 091$ (per day)

$$\text{HP} = \frac{8''.794}{\Delta}$$

$$\text{Equatorial SD} = \frac{98''.47}{\Delta}$$

$$\text{Polar SD} = \frac{91''.91}{\Delta}$$

Once again, compute data for 1978, November 23 at 0^h ET (1978.895 3), using T as above. We find that $\alpha = 8^h\,46^m\,30^s$, $\delta = +18^\circ\,25'\,49''$, $\Delta = 4.815\ 5$ AU, HP = 1″.83, Equatorial SD = 20″.45, Polar SD = 19″.09. (Check, if required, $t = 1\ 198^d.72$, $M = 99^\circ.60$, $E = 102^\circ.31$.)

The values tabulated in the 1978 *AE* are: $\alpha = 8^h\,46^m\,25^s$, $\delta = +18^\circ\,26'\,11''$, $\Delta = 4.815\ 272\ 3$ AU, HP = 1″.83, Polar SD = 19″.09.

The errors are, in this case, minimal: $+5^s$ in RA, +22″ in dec., and for an approximation might be considered to be a very good result. However, caution is advised; do not be misled into thinking that the results for Jupiter will always be so accurate, although they will be more than adequate for finding purposes.

Example 5(e): Saturn.

$T = 1974.023\ 3$

$P = 29.457\ 72$ tropical years

$\left.\begin{aligned} \omega &= 338^\circ.848\ 40\ (+0^\circ.010\ 86) \\ \Omega &= 113^\circ.220\ 17\ (+0^\circ.008\ 73) \\ i &= 2^\circ.490\ 35\ (-0^\circ.000\ 044) \end{aligned}\right\}$ 1950.0

$e = 0.055\ 716\ 5\ (-0.000\ 003\ 47)$, 1950.0

$e^\circ = 3^\circ.186\ 6$

$a = 9.538\ 843$ AU

$n^\circ = 0^\circ.033\ 459\ 9$ (per day)

$$\text{HP} = \frac{8''.794}{\Delta}$$

$$\text{Equatorial SD} = \frac{83''.33}{\Delta}$$

$$\text{Polar SD} = \frac{74''.57}{\Delta}$$

Following the same procedure as in the previous examples, we find, for 1978, November 23 at 0^h ET (1978.895 3): $\alpha = 11^h\,02^m\,50^s$, $\delta = +7^\circ\,57'\,33''$, $\Delta =$

9.470 3 AU, HP = $0''.93$, Equatorial SD = $8''.80$, Polar SD = $7''.87$. (Check, if required, $t = 1\,779^d.45$, $M = 59^\circ.54$, $E = 62^\circ.36$.)

The values tabulated in the 1978 *AE* are: $\alpha = 11^h\,00^m\,00^s$, $\delta = +8^\circ\,13'\,38''$, $\Delta = 9.467\,005\,7$, HP = $0''.93$, Polar SD = $7''.88$.

The planet does not, in fact, reach the computed apparent RA until December 10, and the error in the approximation of RA is just over half a degree, really adequate only for finding purposes. Caution is advised in using these elements, in view of the perturbative forces exerted by the other giant planets.

NOTES

Appendices

APPENDIX I

Visual Binary Star Orbits

The following list is given for those who are unable to refer to Finsen and Worley, *Third Catalogue of Orbits of Visual Binary Stars*, Rep. Obs. Circ. No. 129, Johannesburg, 1970, and whose only working reference is the *Atlas Cæli Catalogue* section which gives a selection of orbits. Many of the orbits listed in that section have been reviewed in the light of more recent observations and the more up-to-date orbits have been included in Finsen and Worley's catalogue.

This list will enable users of the *Atlas Cæli Catalogue* to identify those orbits which are still currently in use. It gives, in addition, the epoch (if published) so that corrections for the effect of precession on the position angle may be made if desired. This does not mean that the earlier orbits are no longer of any use. They can still be employed for computing an ephemeris, by the methods of Chapter 7, but the results will not match those quoted, for example, in the *BAA Handbook* each year. When such orbits are used, the computer should employ the suggested treatment for the correction for precessional effects given in the introduction to Topic 2 of Chapter 7.

ADS	Epoch	Computer	ADS	Epoch	Computer
61	–	Baize	2799	1900	Wierzbinski
161	1900	Arend	2959	–	Couteau
221	–	Muller	2995	1900	Rabe
293	–	Muller	3041	–	Muller
520	–	van den Bos	3082	1950	Muller
755	–	Muller	3159	1900	van den Bos
918	–	Eggen	3182	–	van Biesbroeck
1097	–	Muller	3230	–	Horeschi
1123	–	van den Bos	3248	1950	van den Bos
1158	–	van den Bos	3264	1925	Kuiper
(δ31)	–	Wieth-Knudsen	(h3683)	1900	Wierzbinski
1615	1900	Rabe	3588	–	van den Bos
1630	–	Muller	*3701	–	Eggen
1709	–	Heintz	3841	–	Merrill
2122	1900	Rabe	4229	–	Baize
2200	–	van den Bos	4617	–	Alden
2402	1950	van den Bos	(R65)	–	Eggen
2446	1900	Rabe	(φ19)	–	Finsen
2524	–	Muller	5400	–	Brosche

* Insert 1910.60 for *T* in the *Atlas Coeli Catalogue*.

ADS	Epoch	Computer
5447	1900	Dommanget
5559	1900	Hopmann
5871	1900	Karmel
6175	1900	Rabe
6420	–	Woolley-Symms
6483	–	van den Bos
6549	–	van Biesbroeck
6762	–	Ekenberg
7284	–	van den Bos
7307	–	Arend
7390	1950	Muller
7662	–	Baize
7704	1900	Wierzbinski
7724	1900	Rabe
7744	–	Baize
7780	–	Baize
(φ47)	–	van den Bos
8344	–	Baize
(I 83)	–	van den Bos
8739	–	Baize
8804	1900	Haffner
8891	–	Russell
8939	1900	Baize
8949	1900	Heintz
8974	1900	Wierzbinski
9031	2000	Strand
9182	–	Hopmann
9229	1900	van den Bos
9247	–	Couteau
9301	–	van den Bos
9324	1950	Güntzel-Lingner
9343	1900	Wierzbinski
(φ309)	–	Finsen
(h4707)	–	Woolley-Mason
9425	1900	Heintz
9505	–	Eggen
9557	–	Baize
9617	1900	Danjon
9626	1900	Baize
(h4786)	1900	Heintz
9757	1900	Baize
9769	1950	Giannuzzi
9909	1884	Baize
9932	1910	Wilson
9979	1900	Rabe
10157	1900	Baize
10235	1900	Rabe
10279	1950	Giannuzzi
10360	–	Eggen
10417	–	Brosche
10421	1950	van den Bos
(Brs 13)	2000	Wieth-Knudsen
(Mlb 4)	–	Hirst
10598	–	Duncombe-Ashbrook
11046	2000	Strand
(h5014)	1900	Wierzbinski
11060	–	van Biesbroeck
11324	1900	Heintz
11468	–	Wilson
11483	1900	Heintz
11484	–	Florsch
11520	–	van den Bos
11579	1900	Baize
11635a	1950	Güntzel-Lingner
11635b	1950	Güntzel-Lingner
11989	–	Gottlieb
12096	–	Voronov
12145	1900	Baize
12214	–	van den Bos
(I 253)	1950	van den Bos
12752	–	Güntzel-Lingner
12880	1900	Rabe
(I 120)	1900	Wierzbinski
12973	–	Finsen
13125	–	van Biesbroeck
(HdO 294)	–	van den Bos
13723	1900	Wierzbinski
13728	–	Muller
13850	–	Baize
13944	–	Baize
14073	–	Finsen
14296	1900	Rabe
14499	1900	van den Bos
14773	–	Luyten-Ebbighausen
14775	–	Baize
14783	1900	Baize
14787	1900	van Biesbroeck
15176	–	Danjon
15281	–	Luyten
16046	–	Muller
16057	1900	Heintz
16138	–	Harris
16173	–	Baize
16345	–	Baize
16393	–	Muller
16428	–	Güntzel-Lingner
16497	–	Hirst
16539	–	Muller
16538	–	van Biesbroeck
16666	1900	Wierzbinski
16708	–	van den Bos
17175	2000	Hall

APPENDIX II

Programmes for Calculators Using RPN Logic

The programmes in this Appendix were specially written to suit Hewlett-Packard programmable calculators. That is to say, they were devised by users who had already exercised their choice in favour of RPN logic. As I have indicated earlier in the book, I do not wish to be drawn into any controversy over the respective merits of RPN or algebraic logic and keyboard notation; you should find no difficulty in transposing, if necessary, the Hewlett-Packard RPN instructions of the programmes in this Appendix into a suitable format for your own programmable, whether RPN or algebraic.

All the HP-25 programmes, and 5 of those for the HP-67, were devised or adapted by the author. The majority of the HP-67 programmes were devised by M. Jean Meeus, to whom I am extremely grateful for his many suggestions and kind permission to reproduce them here. The HP-67 programmes are also, of course, fully compatible with the HP-97 desk-top calculator, while with a little readjustment the HP-25 programmes can be run on the HP-19C, HP-29C, HP-55, HP-67 and HP-97.

The programmes are arranged to cover applications in the same subject order as the main text; this group is followed by a collection of useful programmes on subjects which have not been covered elsewhere in the book.

Improvements to existing programmes, and new programmes, are continually being devised. So far as is possible, the range of programmes in the Appendix is up-to-date at the time of going to press. Further improvements, and any new astronomical programmes which are likely to have a wide interest, will be incorporated in later editions.

(1) **HP-25**

To compute a daily ephemeris for Greenwich Mean Sidereal Time at 0^h UT, throughout the year, correct to $\pm 0^s.01$.

---

1. Compute GMST for 0^h UT on January 0 of year, from Topic 1 of Chapter 1. Convert to decimal hours and store in R_0.

2. Enter the programme:

						Register contents:
01	1	07	+	13	GTO 15	R_0 GMST at 0^h UT January 0
02	STO + 1	08	2	14	–	R_1 Day number
03	RCL 1	09	4	15	f H.MS	R_2 Daily rate of gain, ST over MT
04	RCL 2	10	f $x < y$	16	GTO 00	
05	×	11	GTO 14			
06	RCL 0	12	R ↓			

3. Switch to RUN, f PRGM, f FIX 6.
   Enter constant: 0.065 709 822 STO 2
   R/S.

4. The display shows, in H.MS format, the GMST at 0^h UT on January 1.

For the next day, press R/S. The display now shows GMST at 0^h UT on January 2.

For each successive day, press R/S.

5. Test:

Given, for 1978, January 0, $6^h.620\,355\,556$, compute GMST at 0^h UT daily, up to and including January 5.

The results are: January 1 $6^h\,41^m\,09^s.83$
2 $6^h\,45^m\,06^s.39$
3 $6^h\,49^m\,02^s.95$
4 $6^h\,52^m\,59^s.50$
5 $6^h\,56^m\,56^s.05$

2) **HP-25**

To compute a 5-day ephemeris for Greenwich Mean Sidereal Time at 0^h UT, correct to $0^m.1$, and the Julian Date.

This type of ephemeris is published in handbooks such as that issued annually by the BAA. Alongside the date and GMST at 0^h, these ephemerides usually add the Julian Date.

When constructing a 5-day ephemeris, the *second* date to be listed is the midnight following the integral Julian Date nearest to January 0 which is exactly divisible by 5. The *first* date to be listed is that which precedes the second date by 5 days, thus placing this date near the end of the previous year (see the test example).

There are two preparatory calculations to be made before the programme can be run:

a) from Topic 1 of Chapter 1, compute GMST for 0^h UT on January 0 of year and convert to decimal hours;

b) from Programme 32 compute the Julian Date for the same instant (data input at Step 4 is YYYY.00).

. Enter the programme:

						Register contents:
01	5	10	4	19	f last x	R_0 GMST at 0^h UT January 0
02	STO + 1	11	f $x < y$	20	g FRAC	R_1 Day number
03	STO + 3	12	GTO 15	21	RCL 4	R_2 Daily rate of gain, ST over MT
04	RCL 1	13	R ↓	22	×	R_3 JD
05	RCL 2	14	GTO 16	23	f FIX 1	R_4 60
06	×	15	−	24	R/S	
07	RCL 0	16	f FIX 0	25	RCL 3	
08	+	17	f INT	26	GTO 00	
09	2	18	f pause			

. Switch to RUN, f PRGM.

Enter constants: GMST at 0^h UT on January 0, STO 0
0.065 709 822 STO 2
JD at 0^h UT on January 0, STO 3
60 STO 4

. Prepare the registers for the first ephemeris date:
(For 1978) 8 STO − 1, STO − 3 (see test, below).

. *Press R/S.

At line 18 the programme pauses to flash the integral hours of GMST, then continues to line 24 when it will stop to display the minutes, to one decimal place. Press R/S. The programme ends by displaying the Julian Date.

For the next ephemeris date return to * and the programme will output data for 0^h UT five days later.

Repat for each successive 5-day interval.

. Test:

Given GMST for 0^h UT 1978, January 0 is $6^h.620\ 355\ 556$, and the Julian Date at that time is 2 443 508.5, compute the first five entries for a 5-day ephemeris for 1978. The integral JD at 12^h UT on January 0 is thus 2 443 509. This is not exactly divisible by 5. But the next following date is (2443510). The second date in the ephemeris for 1978 is therefore the following midnight, i.e., 0^h UT January 2. It follows that the first date should be 1977, December 28, and this is shown in the ephemeris as 1978, January –3. So, because lines 1 to 3 of the programme add 5 days to the values stored in R_1 and R_3, the Step 3 entry becomes 8 STO – 1, STO – 3.

The programme results are:

1978, January –3	$6^h 25^m.4$	JD = 2 443 505.5
2	$6^h 45^m.1$	510.5
7	$7^h 04^m.8$	515.5
12	$7^h 24^m.5$	520.5
17	$7^h 44^m.2$	525.5

(3) **HP-2**

To compute (for any location) the local mean sidereal time at any time during the year, from local clock time, correct to $\pm 0^s.01$.

1. Enter the programme:

						Register contents:
01	g → H	12	×	23	RCL 7	R_0 Day of month
02	RCL 4	13	RCL 3	24	f $x < y$	R_1 Days to end of previous month
03	+	14	+	25	GTO 28	R_2 Daily rate of gain, ST over MT
04	STO 6	15	RCL 6	26	R ↓	R_3 GMST 0^h UT January
05	RCL 7	16	+	27	GTO 30	R_4 Zone difference in hours
06	÷	17	RCL 5	28	–	R_5 Longitude, D.MS
07	RCL 0	18	g → H	29	GTO 23	R_6 UT
08	RCL 1	19	1	30	f H.MS	R_7 24
09	+	20	5	31	GTO 00	
10	+	21	÷			
11	RCL 2	22	+			

2. Switch to RUN, f PRGM, f FIX 6.
   Store constants:
   0.065 709 822 STO 2
   GMST at 0^h UT on January 0 of year, in decimal hours, from ephemeris c Topic 1 of Chapter 1, STO 3
   *Zone time difference, STO 4 (e.g., EET = –2, CET = –1, Greenwich zon = 0, EST = +5, CST = +6, MST = +7, PST = +8, etc.) Negative E positive W
   Observer's longitude, in D.MS format, positive if E of Greenwich, negative W, STO 5
   24 STO 7
   If British Summer Time (BST) or Daylight Time is in force, key: 1, STO – 4.

3. Enter variables:
   Day of month, STO 0
   Number of days to end of previous month (see table), STO 1
   Local clock time for which mean sidereal time is required, in H.MS form (leave in X register)
   R/S

Number of days to end of previous month

Previous month	Dec	Jan	Feb	Mr	Apr	May	Jun	Jul	Aug	Sep	Oct	Nov
Ordinary year	0	31	59	90	120	151	181	212	243	273	304	334
Leap year	0	31	60	91	121	152	182	213	244	274	305	335

4. Display shows required local mean sidereal time, in H.MS format. For the apparent LST, add the equation of the equinoxes for that date. Many observers choose to ignore the difference between mean and apparent sidereal times.

5. For the LST at a different location return to * in Step 2. For a different date at the same location, return to the beginning of Step 3.

6. Test:

What is the local mean sidereal time at Cambridge, Massachusetts, USA (71° 07′ 30″ W) at 21.30 EST on 1978, May 15?

At Step 2, enter as constants:

R_3= 6.620 355 556
R_4= 5
R_5= 71.073 0 CHS

At Step 3, enter the variables:

R_0= 15
R_1= 120
21.30
R/S

The result is $13^h\ 19^m\ 19^s.45$.

(4) **HP-25**

To compute (for a fixed location) the local mean sidereal time at any time during the year, from local clock time, correct to $\pm 0^s.01$.

When all your LST calculations will be referred to the same geographical location, the previous programme can be amended to simplify the operation so that fewer data inputs are required.

In the programme as written, lines 02 to 04 show three g NOP instructions; when entering the programme, use these lines to accommodate the zone time difference in hours, positive if the site is W of the Greenwich time zone, negative if E (e.g., EET = 2 CHS, CET = 1 CHS, Greenwich zone = 0, EST = 5, CST = 6, MST = 7, PST = 8, etc.). Three spaces are given for this purpose to cater for the case where the entry might be 10 CHS, or similar. If only one or two lines are required, fill the gap with g NOP instructions, so that the entry for line 05 is +.

After line 31 there is more than sufficient programme memory space available to accommodate the longitude of the observer's position, in decimal hours format, positive if E of Greenwich, negative if W (e.g., 80° 22′ 55″.8 W is entered as 5.358 811 CHS). The last line of the programme must be the instruction GTO 22.

The programme includes provision for the amendment necessary when British Summer Time (BST) or Daylight Time (clocks set one hour in advance of mean zone time) is in force. No changes to the day number are required for times near midnight; the programme always works from the observer's clock time and civil date.

Personalize your programme by writing alongside lines 02 to 04, and from 31 onwards, the entries pertinent to your location. Then, when entering the programme on a subsequent occasion, you will not need to stop to remember your exact longitude.

---

1. Enter the programme:

01 g → H	15 +		Register contents:
02 g NOP	16 RCL 2	29 GTO 23	R_0 Day of month
03 g NOP	17 ×	30 f H.MS	R_1 Days to end of
04 g NOP	18 RCL 3	31 GTO 00	previous month
05 +	19 +	32	R_2 Daily rate of gain,
06 RCL 4	20 +	33 Longitude,	ST over MT
07 +	21 GTO 32	34 see	R_3 GMST 0^h UT
08 ENT ↑	22 +	35 note.	January 0
09 ENT ↑	23 RCL 5	36 Last	R_4 0 or –1
10 RCL 5	24 f $x < y$	37 entry	R_5 24
11 ÷	25 GTO 28	38 must be	
12 RCL 0	26 R ↓	39 GTO 22	
13 RCL 1	27 GTO 30		
14 +	28 –		

---

2. Switch to RUN, f PRGM, f FIX 6.

   Store constants: 0.065 709 822 STO 2

   GMST at 0^h UT on January 0 of year, in decimal hours, from ephemeris or Topic 1 of Chapter 1, STO 3

   24 STO 5

   If BST or Daylight Time is in force, key: 1, STO – 4

3. Enter variables: * Day of month STO 0
Number of days to end of previous month (see table), STO 1
† Local clock time, in H.MS format (e.g., 9.35 pm is entered as 21.35) for which LST is required; leave in X register.
R/S

Number of days to end of previous month

Previous month	Dec	Jan	Feb	Mr	Apr	May	Jun	Jul	Aug	Sep	Oct	Nov
Ordinary year	0	31	59	90	120	151	181	212	243	273	304	334
Leap year	0	31	60	91	121	152	182	213	244	274	305	335

4. Display shows the local mean sidereal time, in H.MS format. For the apparent LST, add the equation of the equinoxes.

5. For a different clock time, on the same date, return to † in Step 3.
For a different date, return to * in Step 3.

6. Test:

Assume the observer's longitude is 80° 22′ 55″.8 W, and the year is 1978. The time zone difference is therefore +5 hours. Lines 02 to 04 are entered as 5, g NOP, g NOP.

After line 31, the entries continue: 5.358 811, CHS, GTO 22. Find the local apparent sidereal time at $4^h\,44^m\,30^s$ clock time (EST) on July 7, given GMST at 0^h UT on 1978, January 0 is $6^h.620\,355\,556$ and the equation of the equinoxes is $+0^s.06$.

The variables stored at Step 3 are:

$R_0 = 7$
$R_1 = 181$
4.44 30
R/S

The display gives the local mean sidereal time as $23^h\,22^m\,59^s.98$. Add the equation of the equinoxes to obtain the local apparent sidereal time, which is $23^h\,23^m\,00^s.04$. The example on p 537 of the 1978 *AE* gives $23^h\,23^m\,00^s.044$.

Many workers choose to ignore the difference between mean and apparent sidereal time, unless engaged in high-precision measurements.

In addition to personalizing your programme by writing the zone time difference and longitude against the appropriate programme lines, it is also advisable to calculate, once a year, the entry for R_3 and note this in the margin for easy reference.

(5) **HP-67**

To compute:

(a) the local mean sidereal time (and, if required, the Julian Date) at any instant after 0h UT on 1582, October 15, for any geographical location;

(b) Greenwich Mean Sidereal Time (and Julian Date, if required) at 0h UT on January 0 of any year from 1583;

(c) a daily ephemeris for GMST (and Julian Date) at 0h UT, throughout the year; all to an accuracy of ±0s.001.

Section (c) can be amended to produce an abridged ephemeris (e.g., a 5-day ephemeris, giving hours and minutes of GMST, to one decimal place in the minutes) such as that included in the *BAA Handbook* or similar yearbooks.

1. Load the programme from a magnetic card:

001	f LBL 1	041	6	081	STO 7	121	RCL 1
002	ENT ↑	042	0	082	STO 5	122	f INT
003	f INT	043	0	083	RCL 3	123	STO 1
004	STO 1	044	1	084	STO + 5	124	6
005	−	045	×	085	f INT	125	0
006	EEX	046	f INT	086	STO + 7	126	STO 9
007	2	047	+	087	h last x	127	g x^2
008	×	048	1	088	g FRAC	128	RCL C
009	ENT ↑	049	7	089	RCL D	129	×
010	f INT	050	2	090	÷	130	STO 6
011	STO 2	051	0	091	RCL 7	131	÷
012	−	052	9	092	RCL D	132	h last x
013	EEX	053	9	093	÷	133	h x ←→
014	2	054	5	094	+	134	f INT
015	×	055	+	095	f LBL 6	135	×
016	STO + 3	056	RCL 1	096	STO 4	136	RCL 1
017	RCL 2	057	EEX	097	8	137	h x ←→
018	5	058	2	098	6	138	−
019	f $\sqrt{x}$	059	÷	099	4	139	RCL 6
020	g $x \leq y$	060	f INT	100	0	140	÷
021	GTO 2	061	STO O	101	1	141	RCL C
022	1	062	−	102	8	142	×
023	STO − 1	063	2	103	4	143	6
024	1	064	+	104	.	144	.
025	2	065	RCL O	105	5	145	6
026	STO + 2	066	4	106	4	146	4
027	f LBL 2	067	÷	107	2	147	6
028	RCL D	068	f INT	108	×	148	0
029	EEX	069	+	109	STO 1	149	6
030	2	070	2	110	g FRAC	150	5
031	÷	071	4	111	STO 2	151	5
032	RCL 1	072	1	112	.	152	6
033	×	073	5	113	0	153	+
034	f INT	074	0	114	9	154	h RTN
035	RCL 2	075	2	115	2	155	f LBL A
036	1	076	0	116	9	156	R/S
037	+	077	.	117	RCL 4	157	f H ←
038	3	078	5	118	g x^2	158	RCL B
039	0	079	STO E	119	×	159	+
040	.	080	−	120	STO + 2	160	STO 8

161	RCL C	177	GTO 4	193	g FRAC	209	h R ↓
162	÷	178	h R ↓	194	RCL 9	210	.
163	STO 3	179	GTO 5	195	×	211	0
164	h R ↓	180	f LBL 4	196	f INT	212	1
165	f GSB 1	181	−	197	f − x −	213	+
166	RCL A	182	GTO 3	198	h last *x*	214	f GSB 1
167	f H ←	183	f LBL 5	199	g FRAC	215	GTO 3
168	1	184	f $x > 0$	200	RCL 9	216	f LBL C
169	5	185	GTO 7	201	×	217	1
170	÷	186	RCL C	202	RCL 2	218	STO + 7
171	+	187	+	203	+	219	STO + 5
172	RCL 8	188	f LBL 7	204	DSP 3	220	RCL 7
173	+	189	DSP 0	205	h RTN	221	RCL D
174	f LBL 3	190	f INT	206	f LBL B	222	÷
175	RCL C	191	f − x −	207	0	223	f GSB 6
176	g $x \leq y$	192	h last *x*	208	STO 3	224	GTO 3

2. Before running the programme, store the following data:
   (*i*) longitude of site, in D.MS format, positive if E of Greenwich, negative if W, STO A
   (*ii*) zone time difference, in hours, negative if E of Greenwich, positive if W, STO B
   (e.g., EET = 2 CHS, CET = 1 CHS, Greenwich zone = 0, EST = 5, CST = 6, MST = 7, PST = 8, etc.)
   (*iii*) 24 STO C
   (*iv*) 36 525 STO D

If British Summer Time (BST) or Daylight Time is in force, key: RCL B, 1, −, STO B.

3.(a) **For local mean sidereal time:**

Enter date, in YYYY.MMDD format (e.g., 1978, March 2 is entered as 1978.03 02).

Press A.

Enter clock time for which LST is required, in H.MS format (e.g., 9.30 pm is entered as 21.30).

Press R/S.

* The display will flash the integral hours of LST (line 191). Next, it will flash the integral minutes (line 197). Finally the programme will end by displaying the seconds of LST, to 3 decimal places.

For the apparent local sidereal time, add the equation of the equinoxes.

For the Julian Date: RCL 5, RCL E, f INT, +

(b) **For GMST at 0^h UT on January 0:**

Enter the year (e.g., 1978).

Press B.

The programme will give the same sequence of data outputs as from * in (a) above, but showing GMST instead of LST. For the Julian Date: RCL 5, RCL E, f INT, +

(c) **For a daily ephemeris for GMST at 0^h UT:**

First do Step 3(b) above.

Note the result for January 0 of year plus, if required, the Julian Date.
Press C.
The programme outputs data for January 1.
For each successive day, press C.

(d) If an abridged ephemeris for GMST at 0^h UT must be constructed (e.g., at 5-day intervals, giving hours and minutes of GMST, to one decimal place in the minutes) amend the programme as follows:

Switch to W/PRGM
GTO .197 (Display reads 197 31 84)
h DEL
h DEL
DSP 1
h RTN
SST (Display reads 198 35 82)
GTO .217 (Display reads 217 01)
h DEL (Display reads 216 31 25 13)
5 (Display reads 217 05)
GTO .224 (Check display reads 224 22 03)
Switch to RUN.

To put the programme back into its original form, simply re-load the magnetic card into the programme memory. When constructing a 5-day ephemeris, take as the *second* date the midnight following the integral Julian Date nearest to January 0 which is exactly divisible by 5. In the case of 1978, for example, the integral Julian Date at 12^h on January 0 is 2 443 509, which is not divisible by 5. But the JD, at 12^h on January 1, is 2 443 510, which *is* divisible by 5. The second ephemeris date should therefore be the following midnight, which is 0^h UT January 2. The first ephemeris date must precede this by 5 days, putting it at the end of the previous year. In the case under consideration this will be 1977, December 28, which should be shown in the ephemeris as 1978, January –3. The amended ephemeris will give GMST at 0^h on every fifth day. Follow the method as demonstrated in the test example 4(d).

---

4. Test examples:

(a) What was the apparent local sidereal time for an observatory at Rainham, Kent, 0° 35′ 54″.4 E, at 12.30 am BST on 1976, September 22, given that the equation of the equinoxes was $+0^s.619$?

Initialize according to Step 2:
0.35 54 4 STO A
0 STO B
24 STO C
36 525 STO D
RCL B, 1 , – , STO B (BST was in operation)
Enter date: 1976.09 22 press A
Enter clock time: 0.30 press R/S
The mean sidereal time is given as $23^h\ 36^m\ 13^s.733$. Add to this $+0^s.619$ to

obtain the apparent LST: $23^h\ 36^m\ 14^s.352$. What was the Julian Date? (RCL 5, RCL E, f INT, +) = 2 443 043.479.

)) Find GMST at 0^h UT on 1978, January 0.

Initialize according to Step 2:

0 STO A, STO B

24 STO C

36 525 STO D

Enter year: 1978 press B

The required GMST is $6^h\ 37^m\ 13^s.280$

:) Compute a daily ephemeris for GMST at 0^h UT for 1978, up to and including anuary 5.

Carry out example (b) above.

GMST for January 0 (at 0^h UT) is $6^h\ 37^m\ 13^s.280$.

Press C.

On January 1 it is $6^h\ 41^m\ 09^s.836$. Continue to press C for each successive day. he values obtained are:

January 2	$6^h\ 45^m\ 06^s.391$
3	$6^h\ 49^m\ 02^s.946$
4	$6^h\ 52^m\ 59^s.502$
5	$6^h\ 56^m\ 56^s.057$

What was the Julian Date on January 5? Press: RCL 5, RCL E, f INT, + . The ılian Date is 2 443 513.5.

l) Compute the first five entries for a 5-day ephemeris for GMST at 0^h UT, for ʔ78, giving the data in hours and minutes (to one decimal place), plus the Julian ates.

We have already established at Step 3(d) that the ephemeris should start on anuary –3.

Amend the programme according to Step 3(d).

Initialize according to Step 2:

0 STO A, STO B

24 STO C

36 525 STO D

Enter year: 1978

Press B. This gives GMST for 0^h UT on January 0 (6^h 37m.2). We do not need ıis result, but the process has placed basic data in R_5 and R_7.

*Key: 8, STO – 5, STO – 7. Press C.

The display gives GMST for 0^h UT on 1978, January –3, and the Julian Date can retrieved in the usual manner. Press C for each successive ephemeris entry, taining the following data outputs:

Vhy 8? Because the data in R_5 and R_7 relate to January 0. On pressing C the calculator ll add 5 (line 217) and give data for January 5, which we do not want. So, we deduct 8 om R_5 and R_7, the calculator adds 5 at line 217, and the data output is for January 0 8 +5 = January –3.

1978, January –3	$6^h 25^m.4$	JD ÷ 2 443 505.5
2	$6^h 45^m.1$	510.5
7	$7^h 04^m.8$	515.5
12	$7^h 24^m.5$	520.5
17	$7^h 44^m.2$	525.5

The results agree exactly with the values given in the 1978 *BAA Handbook.*

Note: Because of lines 202 and 203, it is possible on very rare occasions for the seconds display to exceed 60 at the end of a programme run (e.g., for 0^h UT at Greenwich on 1900, January 31, when the programme will give the GST as $8^h 38^m 60^s.775$). If this should occur, simply deduct 60 from the seconds display and add 1 to the integral minutes (e.g., in the example quoted, the GST is $8^h 39^m 00^s.775$).

HP-6

(6)

To compute values for ζ_0, z, θ, $\sin\theta$, and $\tan\frac{1}{2}\theta$ at the beginning of any Besselian sola year, for use in the reduction of the mean place of a star from a standard (catalogue epoch (t_0) to the desired year (t), or vice versa.

Use this programme only when it is required to know the values of the preces sional constants, e.g., as data for inclusion in an ephemeris, or when the constant are not for immediate use in specific calculations. When actual reductions of mea places are to be performed, Programme 11 will be much more convenient to use because it combines the evaluation of the constants *and* the reduction, all in th same programme.

The present programme includes secular changes in the coefficients in terms of τ^2 (after Professor Eichhorn) so the results may differ slightly from those publishe in the *AE*.

---

1. Load the programme from a magnetic card:

001	f LBL A	017	3	033	2	049	×
002	DSP 9	018	÷	034	3	050	+
003	R/S	019	STO 2	035	0	051	.
004	STO 4	020	g x^2	036	4	052	0
005	–	021	STO 3	037	2	053	6
006	EEX	022	RCL 1	038	.	054	RCL 3
007	3	023	g x^2	039	5	055	×
008	÷	024	STO 4	040	3	056	+
009	STO 1	025	RCL 1	041	ENT ↑	057	RCL 1
010	RCL 4	026	×	042	1	058	×
011	1	027	STO 5	043	3	059	3
012	9	028	3	044	9	060	0
013	0	029	6	045	.	061	.
014	0	030	0	046	7	062	2
015	–	031	0	047	3	063	3
016	EEX	032	STO 6	048	RCL 2	064	ENT ↑

065	.	098	+	131	RCL 1	164	R/S
066	2	099	.	132	×	165	RCL 1
067	7	100	3	133	4	166	R/S
068	RCL 2	101	2	134	2	167	RCL 2
069	×	102	RCL 5	135	.	168	R/S
070	–	103	×	136	6	169	f sin
071	RCL 4	104	+	137	7	170	R/S
072	×	105	RCL 6	138	CHS	171	RCL 2
073	+	106	÷	139	ENT ↑	172	2
074	1	107	STO 7	140	.	173	÷
075	8	108	2	141	3	174	f tan
076	RCL 5	109	0	142	7	175	h RTN
077	×	110	0	143	RCL 2	176	f LBL C
078	+	111	4	144	×	177	RCL 1
079	STO 7	112	6	145	–	178	CHS
080	RCL 6	113	.	146	RCL 4	179	R/S
081	÷	114	8	147	×	180	RCL 0
082	STO 0	115	5	148	+	181	CHS
083	RCL 7	116	ENT ↑	149	4	182	R/S
084	7	117	8	150	1	183	RCL 2
085	9	118	5	151	.	184	CHS
086	.	119	.	152	8	185	R/S
087	2	120	3	153	RCL 5	186	f sin
088	7	121	3	154	×	187	R/S
089	ENT ↑	122	RCL 2	155	–	188	RCL 2
090	.	123	×	156	RCL 6	189	CHS
091	6	124	–	157	÷	190	2
092	6	125	.	158	STO 2	191	÷
093	RCL 2	126	3	159	RCL 7	192	f tan
094	×	127	7	160	STO 1	193	h RTN
095	+	128	RCL 3	161	h RTN		
096	RCL 4	129	×	162	f LBL B		
097	×	130	–	163	RCL 0		

2. Enter t (e.g., 1978.0), press A
   Enter t_0 (e.g., 1950.0), press R/S

3. If reductions are to go $t_0 \rightarrow t$, press B
   Note ζ_0; press R/S
   Note z; press R/S
   Note θ; press R/S
   Note $\sin\theta$; press R/S
   Note $\tan\frac{1}{2}\theta$
   For new case, return to Step 2

4. If reductions are to go $t \rightarrow t_0$, press C
   Note ζ_0; press R/S
   Note z; press R/S
   Note θ; press R/S
   Note $\sin\theta$; press R/S
   Note $\tan\frac{1}{2}\theta$
   For new case, return to Step 2

(7) **HP-25**

To update the coordinates for 1920.0 quoted in the Dover paperback edition of Webb's *Celestial Objects for Common Telescopes* with accuracy sufficient for finding purposes.

1. Enter the programme:

01	g → H	18	f tan	35	RCL 4
02	STO 2	19	×	36	×
03	R ↓	20	RCL 5	37	f FIX 1
04	f INT	21	×	38	R/S
05	f last *x*	22	RCL 6	39	RCL 1
06	g FRAC	23	+	40	f cos
07	EEX	24	RCL 0	41	RCL 5
08	2	25	×	42	×
09	×	26	RCL 1	43	RCL 0
10	RCL 4	27	+	44	×
11	÷	28	RCL 3	45	RCL 2
12	+	29	÷	46	+
13	RCL 3	30	f FIX 0	47	f H.MS
14	×	31	f INT	48	f FIX 2
15	STO 1	32	f pause	49	GTO OO
16	f sin	33	f last *x*		
17	RCL 2	34	g FRAC		

Register contents:
R_0 t (years)
R_1 δ_0 (degrees)
R_2 α_0 (degrees)
R_3 15
R_4 60
R_5 0.005 5
R_6 0.012 9
R_7

2. Switch to RUN, f PRGM.
   Store constants: 15 STO 3
   60 STO 4
   0.005 5 STO 5
   0.012 9 STO 6

3. Enter variables:
   t = year – 1920, STO 0 (e.g., to update to 1950.0, store 30 in R_0)
   *α_0 (in H.M. format) ENT ↑
   δ_0 (in D.M. format), CHS if Southern dec. and leave in X register } see test example
   R/S

4. The display flashes the integral hours of α (line 32) and then stops (line 38) to show the minutes to one decimal place. If $\alpha_0 \approx 23^h\ 59^m$ and the display at line 32 is 24, interpret this as 0^h.
   Press R/S.
   The programme stops to show degrees and minutes of δ.

5. For the next star in the batch to be processed, return to * in Step 3.

6. Test:
   In preparation for an observation session, a batch of stars has been selected and their 1920.0 coordinates are to be updated to 1950.0. The first star on the observing list is η Per. The 1920.0 coordinates are quoted as $\alpha = 2^h\ 44^m.8$, $\delta = +55°\ 34'$.
   Initialize according to Step 2.
   At Step 3, enter the variables:
   30 STO 0
   2.448 ENT ↑

55.34
R/S
The display flashes 2^h, then stops to display $47^m.0$.
Press R/S.
The programme ends by displaying 55° 41′.

Note that when entering α there is a slight difference from the usual method of entering a time in the H.MS format. The digit 8 is actually the first decimal place of the minutes, *not* 80 or 8 seconds. We cannot employ a second decimal point in the display to make this more evident, but the programme sorts it out between lines 04 and 12.

Performing rigorous reductions for precession and proper motion with the HP-25

Method B of Topic 3, Chapter 3, is useful for single reductions, and gives very accurate results. When a batch of similar reductions has to be processed, it is better to automate the computation by means of the programme memory in order to avoid operator keying errors.

The capacity of the HP-25 programme memory is 49 lines and this, unfortunately, is not sufficient to be able to accommodate all the instructions in one pass. When using this particular calculator, therefore, it is necessary to run two programmes, one after the other, noting the intermediate results. If a proper system of working is adopted, no difficulty will be encountered. The method is described here.

1. Prepare A4 batch sheets with the layout shown in the diagram. The best way is to prepare a master copy and have the working blanks photocopied from this.

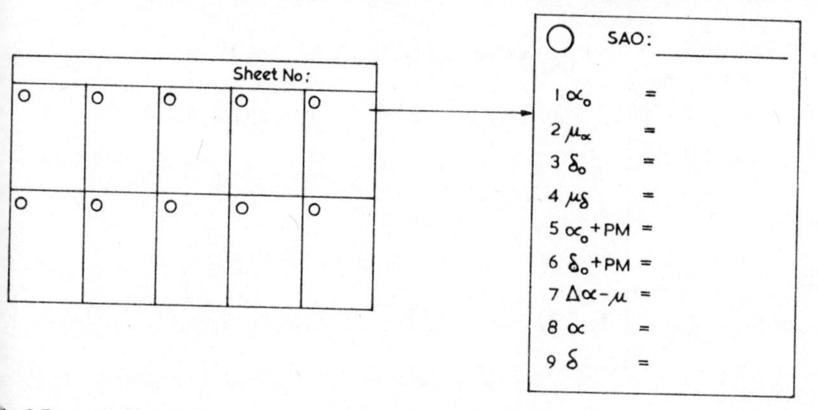

Fig 3 Layout of batch sheet

Each sheet can thus accommodate ten reductions.

2. Take enough sheets to complete the batch and number them serially. On Sheet 1, number the reductions from 1 to 10 in the circles, Sheet 2 from 11 to 20, and so on.

Next, enter the *SAO* number (or the name) of each star, and complete items 1 to 4 from the star catalogue. α_0 is written in H.MS format, δ_0 in D.MS format, ready for the calculator input. Proper motion in RA is entered in seconds of time per annum, while proper motion in dec. is entered in seconds of arc per annum. Be careful with the signs.

3. By Method B of Topic 1, Chapter 2, compute values for ζ_0, z, θ, $\sin\theta$ and $\tan\frac{1}{2}\theta$.

4. Enter Programme 8 into the programme memory, input items 1 to 4 from the batch sheets as instructed, for each star in turn, and record the outputs against items 5 to 7.

5. When all the stars have been processed by Programme 8, clear the programme memory and enter Programme 9. Now, using items 5 to 7 as inputs, process each star in turn once again. Record the outputs against items 8 and 9.

6. This completes the task. Note that item 7 ($\Delta\alpha - \mu$) can be used for updating the proper motions if desired; see Method 4 of Chapter 5 and Programme 10.

○ SAO: 000308

1 α_0	=	1·4848786
2 μ_α	=	0·1811
3 δ_0	=	89·014374
4 μ_δ	=	−0·004
5 α_0+PM	=	27·22440335
6 δ_0+PM	=	89·02878556
7 $\Delta\alpha-\mu$	=	4·923118628
8 α	=	2h 10m 01s·46
9 δ	=	89° 09′ 50″·7

Fig 4 Completed panel after reducing the 1950.0 coordinates for Polaris to 1978.0

(8) HP-25

First of two programmes for the rigorous reduction of equatorial coordinates α, δ from one epoch to another.

1. Read the guidance notes on the preceding pages.

2. Enter the programme:

						Register contents:
01	RCL 5	16	1	31	×	R_0 α_0 (H.MS);
02	×	17	5	32	STO 1	later, $\alpha_0 + \zeta_0$
03	RCL 3	18	×	33	RCL 0	R_1 μ_α;
04	÷	19	R/S	34	f sin	later, q.
05	RCL 2	20	RCL 4	35	×	R_2 δ_0 (D.MS)
06	g→H	21	+	36	RCL 0	R_3 3 600
07	+	22	STO O	37	f cos	R_4 ζ_0
08	RCL 1	23	f cos	38	RCL 1	R_5 $t - t_0$
09	RCL 5	24	RCL 7	39	×	R_6 $\sin\theta$
10	×	25	×	40	CHS	R_7 $\tan\frac{1}{2}\theta$
11	RCL 3	26	$x \longleftrightarrow y$	41	1	
12	÷	27	R/S	42	+	
13	RCL O	28	f tan	43	÷	
14	g→H	29	+	44	g $\tan^{-1}$	
15	+	30	RCL 6	45	GTO OO	

3. Switch to RUN, f PRGM, f FIX 9.
   Enter constants: 3 600 STO 3
   ζ_0 STO 4
   $t - t_0$ STO 5 (where t_0 is the epoch of the catalogue (or known) coordinates and t is the epoch for the required coordinates. If the reduction goes backward in time, this value will be negative)
   $\sin\theta$ STO 6
   $\tan\frac{1}{2}\theta$ STO 7

4. Enter the variables from the batch sheet:
   α_0, in H.MS format, STO 0
   μ_α, in seconds, STO 1
   δ_0, in D.MS format (CHS if Southern dec.) STO 2
   μ_δ, in seconds, leave in X register

Caution: Take care to ensure that μ_α, δ_0 and μ_δ are entered with the correct sign.

5. Press R/S.
   The programme stops at line 19 to display α_0+ PM. Complete item 5 on the batch sheet.
   Press R/S.
   The programme stops again at line 27 to display δ_0+ PM. Complete item 6 on the batch sheet.
   Press R/S.
   The programme concludes by displaying $\Delta\alpha - \mu$. Complete item 7 on the batch sheet.

6. For the next star, return to Step 4.
   Repeat the process until all the stars in the batch have been processed, and all the batch-sheet entries have been completed down to item 7.

7. *Do not switch off the calculator.*

8. Switch to PRGM. Key f PRGM to clear Programme 8 from the programme memory. Do *not* clear the registers, because we have data in R_4, R_3 and R_7 which must be maintained.

9. Proceed to Programme 9.

(9) **HP-25**

Second of two programmes, to complete the rigorous reduction of equatorial co-ordinates α, δ from one epoch to another.

1. Following on from Step 9 of Programme 8, enter the programme:

01 STO 2	16 R ↓	31 CHS	Register contents:
02 RCL 4	17 GTO 19	32 RCL 3	R_0 $\alpha_0 + PM$
03 RCL 5	18 −	33 f cos	R_1 $\delta_0 + PM$
04 +	19 f FIX 6	34 +	R_2 $\Delta\alpha - \mu$
05 +	20 R/S	35 RCL 7	R_3 $\alpha_0 + \zeta_0$
06 RCL O	21 RCL 4	36 ×	R_4 ζ_0
07 +	22 RCL O	37 g $\tan^{-1}$	R_5 z
08 1	23 +	38 2	R_6 $\sin\theta$ (not used)
09 5	24 STO 3	39 ×	R_7 $\tan\frac{1}{2}\theta$
10 ÷	25 f sin	40 RCL 1	
11 f H.MS	26 RCL 2	41 +	
12 2	27 2	42 f H.MS	
13 4	28 ÷	43 f FIX 5	
14 f $x < y$	29 f tan	44 GTO OO	
15 GTO 18	30 ×		

2. Switch to RUN, f PRGM.
Enter constant: z, STO 5.

3. Enter the variables from the batch sheet, for each star in turn:
α_0 + PM, STO 0
δ_0 + PM, STO 1
$\Delta\alpha - \mu$, leave in X register.
R/S.

4. The programme stops at line 20 to display α at the required epoch, in H.MS format. Complete item 8 on the batch sheet. Press R/S. The programme ends by displaying δ, in D.MS format. Complete item 9.

5. For the next star, return to Step 3.
Repeat until all the stars in the batch have been processed.
Take care when entering δ_0 + PM and $\Delta\alpha - \mu$ to ensure that the signs are correct.

6. If revised proper motions are required for the new epoch, proceed to Programme 10. Otherwise, the reductions are now complete.

(10) **HP-25**

To compute values for the proper motions, μ_α' and μ_δ', at the new epoch, after running Programmes 8 and 9.

1. Enter the programme:

01	g → H	14	f sin	27	f cos	Register contents:
02	STO 4	15	×	28	×	R_0 $\mu_{\alpha 0}$
03	RCL 0	16	+	29	RCL 3	R_1 $\mu_{\delta 0}$
04	RCL 2	17	RCL 4	30	f sin	R_2 δ_0 + PM
05	f cos	18	f cos	31	×	R_3 $\Delta\alpha - \mu$
06	×	19	÷	32	RCL 1	R_4 δ'
07	RCL 3	20	f FIX 4	33	RCL 3	R_5 15
08	f cos	21	R/S	34	f cos	
09	×	22	RCL 0	35	×	
10	RCL 1	23	CHS	36	+	
11	RCL 5	24	RCL 5	37	f FIX 3	
12	÷	25	×	38	GTO OO	
13	RCL 3	26	RCL 2			

2. Switch to RUN, f PRGM.
   Enter the constant: 15, STO 5.

3. Enter the variables from the batch sheet, for each star in turn:
   $\mu_{\alpha 0}$, STO 0 (item 2)
   $\mu_{\delta 0}$, STO 1 (item 4)
   δ_0 + PM, STO 2 (item 6)
   $\Delta\alpha - \mu$, STO 3 (item 7)
   δ' in D.MS format (item 9), leave in X register.
   Take care when entering the variables to ensure they bear the correct sign.

4. Press R/S. The programme stops at line 21 to display μ_α' for the required epoch, in seconds of time.
   Press R/S. The programme ends by displaying μ_δ', in seconds of arc.

5. For the next star, return to Step 3.

6. Test:
   Use as data inputs the values shown on the example panel from a batch sheet (section 7 of the explanatory notes before Programme 8).
   Run the programme. μ_α' at 1978.0 is given as $+0^s.208\,1$. μ_δ' at 1978.0 is given as -0″.008.
   If, now, the coordinates and proper motions for 1978.0 are reduced to 1950.0 by Programmes 8 and 9, we should find them in agreement with the original catalogue values. If you try this, remember to use $-z$ for ζ_0, $-\zeta_0$ for z, and to change the signs for $\sin\theta$ and $\tan\frac{1}{2}\theta$ (see Chapter 2, Topic 1).

Rigorous reduction for precession and proper motion from one epoch to another.

---

1. Load the programme from a magnetic card:

```
001 f LBL A
002 f GSB 0
003 GTO 6
004 f LBL B
005 f GSB 2
006 f LBL 6
007 STO A
008 h RTN
009 f LBL C
010 f H ←
011 1
012 5
013 ×
014 STO 1
015 R/S
016 f H ←
017 STO 2
018 R/S
019 2
020 .
021 4
022 ÷
023 STO 3
024 R/S
025 3
026 6
027 ÷
028 STO 4
029 h RTN
030 g LBL a
031 f GSB 0
032 GTO 7
033 g LBL b
034 f GSB 2
035 f LBL 7
036 RCL A
037 –
038 STO B
039 RCL 3
040 ×
041 RCL 1
042 +
043 STO C
044 RCL B
045 RCL 4
046 ×
047 RCL 2
048 +
049 STO D
050 f P ←→ S
051 RCL B
052 RCL 3
053 ×
054 RCL 2
055 +
056 RCL B
057 ×
058 RCL A
059 RCL 1
060 ×
061 RCL 0
062 +
063 +
064 RCL B
065 ×
066 STO E
067 RCL B
068 RCL 9
069 ÷
070 RCL 4
071 +
072 RCL B
073 g x²
074 ×
075 +
076 f P ←→ S
077 STO 8
078 f P ←→ S
079 RCL 5
080 RCL 6
081 RCL A
082 ×
083 –
084 RCL B
085 RCL 7
086 ×
087 –
088 RCL B
089 g x²
090 RCL 8
091 ×
092 –
093 RCL B
094 ×
095 STO B
096 f P ←→ S
097 RCL C
098 RCL E
099 +
100 f cos
101 RCL D
102 f cos
103 ×
104 STO 9
105 RCL B
106 f cos
107 ×
108 RCL D
109 f sin
110 RCL B
111 f sin
112 ×
113 –
114 RCL C
115 RCL E
116 +
117 f sin
118 RCL D
119 f cos
120 ×
121 h x ←→ y
122 g → P
123 h R ↓
124 RCL 8
125 +
126 f x > 0
127 GTO 5
128 3
129 6
130 0
131 +
132 f LBL 5
133 1
134 5
135 ÷
136 g → H.MS
137 f – x –
138 RCL 9
139 RCL B
140 f sin
141 ×
142 RCL D
143 f sin
144 RCL B
145 f cos
146 ×
147 +
148 g sin⁻¹
149 g → H.MS
150 h RTN
151 f LBL 0
152 EEX
153 2
154 ÷
155 1
156 9
157 –
158 h RTN
159 f LBL 2
160 RCL 0
161 g x > y
162 h SF 2
163 h x ←→ y
164 ENT ↑
165 f INT
166 STO 8
167 –
168 EEX
169 2
170 ×
171 ENT ↑
172 f INT
173 STO 9
174 –
175 EEX
176 2
177 ×
178 STO C
179 RCL 9
180 5
181 f √x
182 g x ≤ y
183 GTO 3
184 1
185 STO – 8
186 1
187 2
188 STO + 9
189 f LBL 3
190 RCL 5
191 RCL 8
192 ×
193 f INT
194 RCL 9
195 1
196 +
197 RCL 6
198 ×
199 f INT
200 +
201 RCL C
202 +
203 h F ? 2
204 GTO 4
```

205	RCL 8	210	STO C	215	4	220	h RC I
206	EEX	211	–	216	÷	221	–
207	2	212	2	217	f INT	222	RCL 7
208	÷	213	+	218	+	223	÷
209	f INT	214	RCL C	219	f LBL 4	224	h RTN

2. Load the registers from the magnetic card bearing the data. See Programme 11a, Data input card for Programme 11.

3. Enter the initial (known) epoch:
   either in years and decimals (e.g., 1950.0); press A
   or as a calendar date (in YYYY.MMDD format); press B

4. Enter the variables:
   α, in H.MS format, press C
   δ, in D.MS format, press R/S
   μ_α, in seconds per year, press R/S
   μ_δ, in seconds of arc per year, press R/S

5. Enter the final (required) epoch:
   either in years and decimals; press f a
   or as a calendar date; press f b

6. The programme will pause at line 137 to flash α at the required epoch, in H.MS format, and will continue by computing δ at the required epoch, stopping at line 150 to display this in D.MS format.

7. For revised coordinates for the same star, but at a different new epoch, return to Step 4.

8. For a new star, return to Step 4 if the initial epoch is the same, otherwise to Step 3.

Note: The coordinates given for the new epoch are for the mean place.

Two cards are needed to run this programme, one with the programme instructions and the other bearing data. They may be loaded in either order—the calculator will recognize which is which.

If either of the epochs is entered as a calendar date in YYYY.MMDD format, this date must be later than March 1 of the year 0.

Test: The equatorial coordinates for Polaris at 1950.0 are:

$\alpha = 1^h\,48^m\,48^s.786$

$\delta = +89°\,01'\,43''.74$

$\mu_\alpha = +0^s.181\,1$

$\mu_\delta = -0''.004$

Find the coordinates for epoch 1978.0.

Run the programme and find:

$\alpha = 2^h\,10^m\,01^s.46$ $\quad\quad$ $\delta = +89°\,09'\,50''.7$

Display: When recording the magnetic card, set the display for 6 decimal places.

(11a) **HP-67**

Data input card for Programme 11.

1. Clear all primary and secondary registers, and enter the following data into the calculator:

   1 582.101 5 STO 0
   365.25 STO 5
   30.600 1 STO 6
   36 524.219 9 STO 7
   694 025.813 STO I
   f P ←→ S (for secondary registers)
   0.640 069 444 STO 0
   0.000 387 778 STO 1
   0.000 083 889 STO 2
   0.000 005 STO 3
   0.000 219 722 STO 4
   0.556 856 111 STO 5
   0.000 236 944 STO 6
   0.000 118 333 STO 7
   0.000 011 667 STO 8
   3 600 000 STO 9
   f P ←→ S

2. Press f W/DATA. The calculator displays 'Crd'.
   Pass a blank, unclipped magnetic card through the card reader to record data for the primary registers. Pass the other side of the card through to record data for the secondary registers.

3. Retain this card for use with Programme 11.

(12) **HP-25**

First of two programmes applying rotational geometry to reductions for precession. See Chapter 3, Topic 4.

Three successive rotations are performed on the rectangular equatorial coordinates, x, y, z:

(*i*) about the z_0 axis through angle ζ_0
(*ii*) about the y' axis through angle θ
(*iii*) about the z^* axis through angle $-z$

As written, this programme commences with data inputs in terms of α and δ, followed by adjustment in respect of proper motion in the interval $t - t_0$, after which α, δ are converted into x, y, z. Similarly, at the end of Programme 13, x, y, z are converted back into α, δ.

Where the coordinates are already catalogued in terms of x, y, z, those parts of the two programmes which perform the coordinate conversions can be edited out, provided that satisfactory allowance for the effects of proper motion can be incorporated in the computation.

An orderly system of documentation is advised, perhaps based on that suggested in the preliminary notes for Programmes 8 and 9.

1. Enter the programme:

01 RCL 1	17 RCL 2	33 ×	
02 RCL 7	18 +	34 STO 3	
03 ×	19 f cos	35 RCL 5	Register contents:
04 RCL 6	20 ENT ↑	36 RCL 1	R_0 α_0
05 ÷	21 ENT ↑	37 f → R	later, α_0 + PM
06 RCL 0	22 f last x	38 RCL 5	R_1 μ_α
07 +	23 f sin	39 RCL 3	later, x_0
08 1	24 STO 4	40 f → R	R_2 δ_0
09 5	25 R ↓	41 R ↓	R_3 μ_δ
10 ×	26 RCL 0	42 −	later, y_0
11 STO 0	27 f cos	43 R/S	R_4 z'
12 RCL 3	28 ×	44 R ↓	R_5 ζ_0
13 RCL 7	29 STO 1	45 +	R_6 3 600
14 ×	30 R ↓	46 R/S	R_7 $t - t_0$
15 RCL 6	31 $x \leftarrow\rightarrow y$	47 RCL 4	
16 ÷	32 f sin	48 GTO 00	

2. Switch to RUN, f PRGM, f FIX 9.

3. Enter constants:
   ζ_0, STO 5
   3 600, STO 6
   $t - t_0$, STO 7 (final epoch minus initial epoch)

4. Enter variables:
   α_0, in H.MS format, g → H, STO 0
   μ_α, STO 1 (seconds of time per year)
   δ_0, in D.MS format, g → H, STO 2
   μ_δ, STO 3 (seconds of arc per year)

5. Compute x', press R/S.
   Compute y', press R/S.
   Compute z', press R/S.

6. Record these intermediate results as data inputs for Programme 13, and return to Step 4 for the next star in the batch.

(13) **HP-25**

Second of two programmes, to complete the rotational transformation of rectangular equatorial coordinates from the equator and equinox of one epoch to another.

1. Enter the programme:

						Register contents:
01	RCL 6	16	f→R	31	ENT ↑	R_0 x'
02	RCL 2	17	RCL 7	32	RCL 3	later, x^* and x_1
03	f→R	18	RCL 1	33	÷	R_1 $y' = y^*$
04	RCL 6	19	f→R	34	f H.MS	later, y_1
05	RCL 0	20	R ↓	35	R/S	R_2 z'
06	f→R	21	+	36	$x \leftrightarrow y$	later, $z^* = z_1$
07	R ↓	22	STO 0	37	f sin	R_3 15
08	+	23	R ↓	38	RCL 2	R_4 (not used)
09	STO 2	24	CHS	39	×	R_5 ζ_0
10	R ↓	25	+	40	RCL 1	R_6 θ
11	CHS	26	STO 1	41	÷	R_7 $-z$
12	+	27	RCL 0	42	g $\tan^{-1}$	
13	STO 0	28	÷	43	f H.MS	
14	RCL 7	29	g $\tan^{-1}$	44	GTO 00	
15	$x \leftrightarrow y$	30	ENT ↑			

2. Switch to RUN, f PRGM, f FIX 6.

3. Enter constants:
   15, STO 3
   ζ_0, STO 5
   θ, STO 6
   $-z$, STO 7

4. Enter first rotation coordinates from Programme 12:
   x', STO 0
   y', STO 1
   z', STO 2

5. Compute α for required epoch. Press R/S. The programme stops at line 35 to display α in H.MS format.
   Compute δ for required epoch. Press R/S. The programme ends by displaying δ in D.MS format.

6. Return to Step 4 for the next star in the batch.

7. Test:
   Find the equatorial coordinates for Polaris at 1978.0, given the 1950.0 position and proper motions:

$\alpha = 1^h\ 48^m\ 48^s.786$

$\delta = +89°\ 01'\ 43''.74$

$\mu_\alpha = +0^s.181\ 1$

$\mu_\delta = -0''.004$

From Chapter 2, Topic 1:

$\zeta_0 = 0.179\ 280\ 709$

$z = 0.179\ 297\ 981$

$\theta = 0.155\ 877\ 202$

Run the two programmes, and find, for 1978.0:

$\alpha = 2^{h}\,10^{m}\,01^{s}.46$

$\delta = +89°\,09'\,50''.7$

---

(14) **HP-25**

To iterate for $x - \sin x$.

This programme has been devised for use when calculating the elements of the orbit of a visual binary star; see Chapter 7, Topic 1. It is employed to compute μ by successive iterations for $u - \sin u$, $v - \sin v$, and $(u + v) - \sin(u + v)$.

1. Enter the programme:

01	g RAD	15	RCL 4	29	f pause
02	STO 0	16	2	30	R ↓
03	1	17	÷	31	GTO 05
04	STO 1	18	GTO 20	32	RCL 1
05	ENT ↑	19	RCL 4	33	3
06	f sin	20	STO − 1	34	6
07	−	21	g ABS	35	0
08	RCL 0	22	EEX	36	g π
09	−	23	5	37	2
10	STO 4	24	CHS	38	×
11	2	25	f $x \geqslant y$	39	÷
12	RCL 1	26	GTO 32	40	×
13	f $x < y$	27	RCL 1	41	GTO 00
14	GTO 19	28	RCL 4		

Register contents:
R_0 $x - \sin x$
R_1 Trial x
R_2 μ
R_3 Temporary store
R_4 Trial $x - \sin x$
Multipliers for μ:
R_5 for $u - \sin u$
R_6 for $v - \sin v$
R_7 for $(u + v) - \sin(u + v)$.

2. Switch to RUN, f PRGM, f FIX 4.
   Store constants:
   Multiplier for μ for $u - \sin u$, STO 5 (e.g., for the worked example in Chapter 7, Topic 1, 2.90)
   Multiplier for μ for $v - \sin v$, STO 6 (e.g., 36.28)
   Multiplier for μ for $(u + v) - \sin(u + v)$, STO 7 (e.g., 49.17)
   *Trial μ, STO 2 (e.g., 0.12)

3. To compute μ. Prepare a table as shown in the worked example of Chapter 7, Topic 1.

(a) RCL 2, RCL 5, ×. Note value of $u - \sin u$ on line (*i*) of table. Press R/S. Display gives u in degrees. STO 3. Complete line (*iv*)

(b) RCL 2, RCL 6, ×. Note value of $v - \sin v$ on line (*ii*) of table. Press R/S. Display gives v in degrees. Complete line (*v*). RCL 3, +, STO 3

(c) Complete line (*vi*) for $u + v$

(d) RCL 2, RCL 7, ×. Note value of $(u + v) - \sin(u + v)$ on line (*iii*). Press R/S. Display gives $(u + v)$ in degrees. Complete line (*vii*)

(e) RCL 3, $x \leftarrow\rightarrow y$, −. Complete line (*viii*)

(f) Assess new trial value for μ and return to * in Step 2 for start of next column

(g) Continue the process until entry in line (*viii*) is as close to 0° as can be achieved with no more than 4 decimal places in the value for μ.

Note: Some of the iterative runs are completed rapidly, while others take longer to converge to zero. The converging process can be seen while the programme is running by means of the pause instruction (line 29). When this value falls below 0.000 01 the programme branches to line 39 and converts the current value of x in R_1 to degrees. Values of $x - \sin x$ in excess of 5.0 or less than 1.0 take longest to converge to zero.

4. Test the programme by re-computing the table in Chapter 7, Topic 1. Values computed by means of this programme will differ slightly from those shown in the table; this is because the calculator is working to more decimal places in the registers, although only four are displayed. To this extent, the results are more accurate, but the final value for μ will be the same.

Computing the position angle and separation of a number of double stars with the HP-25

Method B of Topic 2, Chapter 7 is very accurate for use when only a single computation has to be made. When data for more than one binary are required, the work can be simplified and speeded up if the computation can be programmed into the calculator memory. The opportunity for mistakes is greatly reduced because keyboard entries by the operator are kept to a minimum.

However, unlike the HP-67, the capacity of the HP-25 is not sufficient to accommodate all the instructions in one pass, so it becomes necessary to run three programmes consecutively. Data outputs from the first programme have to be noted down in readiness for use as inputs for the second programme, and so on. Just as in the case for rigorous reductions for precession, a proper system of working must be adopted, and a suitable method is described here.

1. Prepare A4 batch sheets, to be used in the horizontal ('landscape') format, with the rulings and headings shown in the diagram. Again, the best way is to prepare a master copy and have the working blanks photocopied from this.

Name 1	Other names 2	P 3	T 4	e 5	a 6	ω 7	i 8	Ω 9
OΣ 547	ADS 48	362·3	1710·0	0·52	6·179	276·58	62·3	19·07
				17mm				

'ig 5a Left side of batch sheet

Epoch 10	α 11	δ 12	E° 13	r 14	θ−Ω 15	v+ω 16	Unc θ 17	Cor θ 18	ρ 19
1950	0·0248	45·32	241·187	7·728	333·774	493·338	352·844	172·8	5·91
		17 mm			13mm				

'ig 5b Right side of batch sheet

Columns 1 to 12 are 17 mm in width, columns 13 to 19 are each 13 mm in width. \ specimen entry shows the data for O$\sum$ 547 and results for 1979.0.

2. Take enough sheets to complete the batch (you should get about 16 stars to the sheet). Complete columns 1 to 12 from a catalogue of binary-star orbits (see the ntroduction to Topic 2 of Chapter 7 for details).

3. Enter and run Programme 15 for all the stars in turn, noting E° and r in columns 13 and 14.

4. Enter and run Programme 16, noting $(\theta - \Omega)$, $(v + \omega)$ and uncorrected θ in columns 15 to 17.

5. Enter and run Programme 17. If an epoch has been given in the catalogue, the programme will give θ corrected for the effect of precession, and will end by displaying the separation, ρ.

(15) HP-2

Position and angle and separation of a visual binary star. To compute E° and r fo Programmes 16 and 17.

1. Prepare a batch sheet as described in the explanatory notes.
   Enter the programme:

						Register contents:
01	RCL 4	18	RCL 7	35	GTO 38	R_0 P; n°; e°
02	g π	19	STO 0	36	RCL 7	R_1 T; M°
03	2	20	R ↓	37	GTO 23	R_2 e
04	×	21	×	38	RCL 7	R_3 a
05	÷	22	STO 1	39	f cos	R_4 360
06	RCL 2	23	STO 6	40	RCL 2	R_5 0.000 1
07	×	24	f sin	41	×	R_6 Trial E
08	STO 7	25	RCL 0	42	CHS	R_7 e; E°
09	RCL 4	26	×	43	1	
10	RCL 0	27	RCL 1	44	+	
11	÷	28	+	45	RCL 3	
12	STO 0	29	STO 7	46	×	
13	R ↓	30	RCL 6	47	RCL 7	
14	R ↓	31	−	48	R/S	
15	RCL 1	32	g ABS	49	$x \leftarrow\rightarrow y$	
16	−	33	RCL 5			
17	RCL 0	34	f $x \geqslant y$			

2. Switch to RUN, f PRGM, f FIX 3.
   Store constants: 360, STO 4
   0.000 1, STO 5

3. Enter variables: P, STO 0
   T, STO 1
   e, STO 2
   a, STO 3
   t, (final epoch), leave in X register

4. Press R/S.
   E° is displayed (if negative, RCL 4, +). Complete column 13. Press R/S.
   The programme finishes by displaying r. Complete column 14.

5. For a new year: P, STO 0, T, STO 1, t (leave in X register).
   (P and T have been lost in the previous run.)
   Press R/S for E and again for r.

6. For a new star, return to Step 3. Repeat for each star in turn until the batc has been completed, then proceed to Programme 16.

(16) **HP-25**

Second of three programmes for computing the position angle and separation of a visual binary star at any required epoch. To compute $(\theta - \Omega)$, $(\nu + \omega)$ and θ.

1. Enter the programme:

						Register contents:
01	ENT ↑	15	g $x \geqslant 0$	29	RCL 6	R_0 $(\nu + \omega)$
02	2	16	GTO 19	30	+	R_1 e
03	÷	17	RCL 3	31	g $x \geqslant 0$	R_2 θ
04	f tan	18	+	32	GTO 35	R_3 180
05	1	19	2	33	RCL 7	R_4 ω
06	RCL 1	20	×	34	+	R_5 i
07	+	21	RCL 4	35	STO 2	R_6 Ω
08	1	22	+	36	RCL 6	R_7 360
09	RCL 1	23	STO 0	37	−	
10	−	24	f tan	38	R/S	
11	÷	25	RCL 5	39	RCL 0	
12	f $\sqrt{x}$	26	f cos	40	R/S	
13	×	27	×	41	RCL 2	
14	g $\tan^{-1}$	28	g $\tan^{-1}$	42	GTO 00	

2. Switch to RUN, f PRGM, f FIX 3.
   Store constants: 180, STO 3
   360, STO 7

3. Enter variables: e, STO 1
   ω, STO 4
   i, STO 5
   Ω, STO 6
   $E°$, (leave in X register)

4. Press R/S.
   $(\theta - \Omega)$ is displayed. Complete column 15.
   Press R/S.
   $(\nu + \omega)$ is now displayed. Complete column 16.
   Press R/S.

The programme ends by displaying θ. If no epoch has been given for the elements (column 10 blank), this is the final value for θ. Complete column 18 to one decimal place. But if an epoch has been given, θ is to be corrected for precession in the next programme; in this case, complete column 17 to 3 decimal places.

5. For a new year: Put a new value of $E°$ in the X register and return to Step 4.

6. For a new star, return to Step 3. Repeat for each star in turn until the batch has been completed, then proceed to Programme 17.

(17) **HP-25**

The last of three programmes for computing the position angle and separation of a visual binary star at any epoch. To compute θ (corrected for precession) and ρ.

1. Enter the programme:

01 g → H		17 RCL 0		33 GTO 40		Register contents:
02 f cos		18 +		34 1		R_0 θ
03 g 1/x		19 STO 0		35 8		R_1 $(\theta - \Omega)$
04 RCL 7		20 RCL 2		36 0		R_2 r
05 g → H		21 RCL 3		37 +		R_3 $(v + \omega)$
06 1		22 f cos		38 GTO 40		R_4 t
07 5		23 ×		39 RCL 0		R_5 epoch or 0
08 ×		24 RCL 1		40 f FIX 1		R_6 0.005 6
09 f sin		25 f cos		41 g $x = 0$		R_7 α H.MS or 0
10 ×		26 g 1/x		42 GTO 45		
11 RCL 4		27 ×		43 f pause		
12 RCL 5		28 g $x \geqslant 0$		44 GTO 46		
13 −		29 GTO 39		45 R/S		
14 ×		30 CHS		46 $x \longleftrightarrow y$		
15 RCL 6		31 RCL 0		47 f FIX 2		
16 ×		32 g $x = 0$		48 GTO 00		

2. Switch to RUN, f PRGM.
Store constant: 0.005 6, STO 6.

3. Enter variables:

(a) If no epoch for the orbit has been given,
$(\theta - \Omega)$, STO 1
r, STO 2
$(v + \omega)$, STO 3
t, STO 4 (final epoch)
0, STO 0, STO 5, STO 7.

(b) If an epoch has been given,
θ, STO 0
$(\theta - \Omega)$, STO 1
r, STO 2
$(v + \omega)$, STO 3
t, STO 4
epoch, STO 5
α (H.MS), STO 7
δ (D.MS), leave in X register

4. Press R/S.

(a) Programme flashes 0 (to signal we already have a final value for θ), and continues automatically, ending by displaying ρ to two decimal places. Complete the final column of the batch sheet.

or (b) Programme stops to display corrected θ to one decimal place. Complete column 18. Press R/S. The programme ends by displaying ρ to two decimal places. Complete the final column of the batch sheet.

5. For a new year: Enter new values for θ, $(\theta - \Omega)$, r, $(v + \omega)$ and t. Re-enter δ (D.MS) in the X register, and return to Step 4.

5. For a new star: Return to Step 3 and repeat until all the stars in the batch have been completed.

This completes the computation.

Test: Use as input data the elements given for O∑ 547 in the example of a batch-sheet layout, in the introductory notes for this set of three programmes.

Run the programmes, and confirm the results for 1979.0.

18) **HP-67**

To compute the position angle and separation of a visual binary star at any epoch.

1. Load the programme from a magnetic card:

001	f LBL A	038	STO 6	075	f LBL 3	112	×
002	h π	039	R/S	076	ENT ↑	113	RCL 9
003	2	040	STO 7	077	f sin	114	h $x \longleftrightarrow y$
004	×	041	f P ←→ S	078	RCL 5	115	STO 9
005	STO 8	042	R/S	079	×	116	h R ↓
006	h $x \longleftrightarrow y$	043	STO 8	080	−	117	2
007	÷	044	R/S	081	RCL 6	118	÷
008	STO 0	045	STO 9	082	−	119	f tan
009	h π	046	f P ←→ S	083	f $x = 0$	120	1
010	f D ←	047	R/S	084	GTO 4	121	RCL 2
011	STO D	048	STO 8	085	h F ? 1	122	+
012	RCL 8	049	f LBL 9	086	GTO 2	123	1
013	f D ←	050	RCL 1	087	RCL 4	124	RCL 2
014	STO E	051	−	088	÷	125	−
015	5	052	RCL 0	089	1	126	÷
016	6	053	×	090	−	127	f $\sqrt{x}$
017	EEX	054	RCL 2	091	h $1/x$	128	×
018	4	055	f P ←→ S	092	RCL 3	129	g $\tan^{-1}$
019	CHS	056	STO 5	093	×	130	f $x > 0$
020	h ST I	057	h R ↓	094	STO − 1	131	GTO 5
021	EEX	058	STO 1	095	h ABS	132	RCL D
022	5	059	STO 6	096	RCL 2	133	+
023	CHS	060	h RAD	097	g $x \leq y$	134	f LBL 5
024	f P ←→ S	061	f LBL 1	098	GTO 1	135	2
025	STO 2	062	h SF 1	099	f LBL 4	136	×
026	f P ←→ S	063	RCL 1	100	RCL 1	137	RCL 4
027	R/S	064	GTO 3	101	f D ←	138	+
028	STO 1	065	f LBL 2	102	h DEG	139	STO A
029	R/S	066	STO 4	103	f P ←→ S	140	f tan
030	STO 2	067	h CF 1	104	STO 9	141	RCL 5
031	R/S	068	RCL 1	105	RCL 3	142	f cos
032	STO 3	069	RCL 1	106	1	143	×
033	R/S	070	EEX	107	RCL 2	144	g $\tan^{-1}$
034	STO 4	071	5	108	RCL 9	145	RCL 6
035	R/S	072	÷	109	f cos	146	+
036	STO 5	073	STO 3	110	×	147	f $x > 0$
037	R/S	074	+	111	−	148	GTO 6

```
149  RCL E
150  +
151  f LBL 6
152  STO B
153  RCL 6
154  -
155  STO C
156  RCL 7
157  f x = 0
158  GTO 7
159  f P ←→ S
160  RCL 9
161  f H ←
162  f cos
163  h 1/x
164  RCL 8
165  f H ←
166  1
167  5
168  ×
169  f sin
170  ×
171  f P ←→ S
172  RCL 8
173  RCL 7
174  -
175  ×
176  h RC I
177  ×
178  f - x -
179  RCL B
180  +
181  STO B
182  f LBL 7
183  RCL 9
184  RCL A
185  f cos
186  ×
187  RCL C
188  f cos
189  h 1/x
190  ×
191  f x > 0
192  GTO 8
193  CHS
194  RCL B
195  RCL D
196  +
197  RCL E
198  g x > y
199  GTO 9
200  -
201  GTO 0
202  f LBL 9
203  h R ↓
204  f LBL 0
205  DSP 1
206  R/S
207  h x ←→ y
208  DSP 2
209  h RTN
210  f LBL 8
211  RCL B
212  GTO 0
213  f LBL B
214  1
215  STO +8
216  RCL 8
217  GTO 9
218  f LBL C
219  STO 8
220  GTO 9
```

2. Enter elements of elliptical orbit:

   P Press A
   T Press R/S
   e Press R/S
   a Press R/S
   ω Press R/S
   i Press R/S
   Ω Press R/S

3. (a) If no epoch for the orbit is given, enter:
   0, R/S, R/S, R/S

   (b) If an epoch has been given, enter:
   epoch (e.g., 1900), R/S
   α in H.MS format, R/S
   δ in D.MS format, R/S

4. Enter t (epoch for which data is required), R/S

5. The programme automatically iterates for E, the eccentric anomaly. If a epoch for the orbit elements has been input, the display will flash (line 178) t show $\Delta\theta$, which is the correction to θ for the effect of precession during the perio $t - t_0$. If no orbit is given, the programme will skip this section.

The programme stops to display θ to one decimal place, corrected, if necessar for precession (i.e., if the display flashes $\Delta\theta$ at line 178 there is no need to note th value down unless it is specifically required for another purpose).

Press R/S. The programme ends by displaying ρ to two decimal places.

6. If constructing an ephemeris for the same star, to find θ, ρ for the next followin year, press B.

7. To find θ, ρ for the same star at any other year, enter the required year an press C.

8. For another star, return to Step 2.

Test: The orbit elements for OΣ 547 are $P = 362.3$, $T = 1\,710.0$, $e = 0.52$, $a = 6.179$, $\omega = 276.58$, $i = 62.3$, $\Omega = 19.07$, the epoch is 1950, $\alpha = 0^h\,02^m\,48^s$, $\delta = +45°\,32'$. Find the position angle and separation for 1979.0 and 1980.0.
Results: 1979.0, $\theta = 172°.8$, $\rho = 5''.91$; 1980.0, $\theta = 173°.3$, $\rho = 5''.92$.

(19) **HP-25**

To iterate for E in Eqn. 8.7 (elliptical orbit).

1. Enter the programme:

						Register contents:
01	STO 6	11	÷	21	CHS	R_0 $M°$
02	RCL 4	12	×	22	f $x \geqslant y$	R_1 Trial E
03	×	13	RCL 0	23	GTO 26	R_2 e
04	STO 0	14	+	24	RCL 5	R_3 180
05	STO 1	15	STO 5	25	GTO 05	R_4 $n°$
06	f sin	16	RCL 1	26	RCL 5	R_5 E
07	RCL 2	17	−	27	GTO 00	R_6 t
08	RCL 3	18	g ABS	28	STO + 6	
09	×	19	EEX	29	RCL 6	
10	g π	20	8	30	GTO 02	

2. Switch to RUN, f PRGM, f REG, f FIX 6
Store constants: e (not $e°$), STO 2 (the programme converts e into $e°$ during lines 05 to 08)
180, STO 3
$n°$, STO 4

3. Enter t (interval in days before or after perihelion, T; before perihelion, t is negative)
Press R/S. E is displayed to six decimal places.

4. For the next date, enter the interval in days into the X register, press GTO 28, R/S. (For example, if a 5-day ephemeris is to be prepared, enter 5, GTO 28, R/S.)

Test: Using data from Example 2 of Chapter 8,
enter $e = 0.145\,446$
$n° = 0.115\,561\,2$
$t = 529.525\,8$ days (for 1977, January 17 at 0^h ET).
Result: $E = 68°.971\,066$
For the next day (January 18) enter 1, GTO 28, R/S, and obtain the result, $E = 69°.092\,972$.

(20) **HP-25**

To iterate for v in Eqn. 8.1 (parabolic orbit).

1. Enter the programme:

			Register contents:
01 STO 4	14 RCL 0	27 CHS	R_0 t × constant
02 RCL 3	15 +	28 f $x \geqslant y$	R_1 Trial v
03 ×	16 RCL 1	29 GTO 32	R_2 v
04 STO 0	17 g x^2	30 RCL 2	R_3 Constant
05 1	18 1	31 GTO 06	R_4 t
06 STO 1	19 +	32 RCL 2	
07 g x^2	20 ÷	33 g $\tan^{-1}$	
08 RCL 1	21 STO 2	34 2	
09 ×	22 RCL 1	35 ×	
10 2	23 −	36 GTO 00	
11 ×	24 g ABS	37 STO + 4	
12 3	25 EEX	38 RCL 4	
13 ÷	26 8	39 GTO 02	

2. Switch to RUN, f PRGM, f FIX 6.
   Store constant: 0.012 163 721, ENT ↑ , q, ENT ↑ , 1.5, f y^x, ÷, STO 3

3. Enter t into X register (interval in days before or after perihelion, T; before perihelion, t is negative)
   Press R/S. v is displayed to six decimal places.

4. For the next date, enter the time interval in days, GTO 37, R/S. (For example, if a 5-day ephemeris is being prepared, enter 5, GTO 37, R/S.)

Test: Using data from Example 1 of Chapter 8,
enter q = 0.218 445
t = 42.826 9 days (for 1976, February 2 at 0^h ET)
Result: v = 128°.737 788
For the next day (February 3) enter 1, GTO 37, R/S and obtain the result, v = 129°.207 663.

To compute geocentric positions for comets with parabolic elements.

1. Load the programme from a magnetic card. When recording the card, set the display for four decimal places.

001	f LBL A	051	h R ↓	101	3	151	×
002	.	052	STO B	102	÷	152	RCL 0
003	0	053	RCL 5	103	RCL 0	153	×
004	1	054	f sin	104	+	154	f P ←→ S
005	2	055	RCL 1	105	RCL 1	155	STO + 2
006	1	056	f sin	106	g x^2	156	f P ←→ S
007	6	057	×	107	1	157	RCL 5
008	3	058	RCL 1	108	+	158	RCL C
009	7	059	f sin	109	÷	159	+
010	2	060	RCL 0	110	STO 5	160	f sin
011	RCL 2	061	×	111	RCL 1	161	RCL E
012	÷	062	RCL 1	112	–	162	×
013	RCL 2	063	f cos	113	h ABS	163	RCL 0
014	f $\sqrt{x}$	064	RCL 3	114	EEX	164	×
015	÷	065	f sin	115	8	165	f P ←→ S
016	h ST I	066	×	116	CHS	166	STO + 3
017	RCL 5	067	+	117	g $x > y$	167	RCL 1
018	f cos	068	g → P	118	GTO 1	168	RCL 2
019	RCL 5	069	STO E	119	RCL 5	169	g → P
020	f sin	070	h R ↓	120	GTO 2	170	g x^2
021	RCL 3	071	STO C	121	f LBL 1	171	RCL 3
022	f cos	072	h RTN	122	RCL 5	172	g x^2
023	×	073	f LBL B	123	g x^2	173	+
024	CHS	074	f P ←→ S	124	1	174	f $\sqrt{x}$
025	g → P	075	STO 1	125	+	175	STO 8
026	STO 6	076	STO 4	126	RCL 2	176	RCL 2
027	h R ↓	077	g x^2	127	×	177	RCL 1
028	STO A	078	STO 7	128	STO 0	178	g → P
029	RCL 5	079	R/S	129	RCL 5	179	h R ↓
030	f sin	080	STO 2	130	g $\tan^{-1}$	180	1
031	RCL 1	081	STO 5	131	2	181	5
032	f cos	082	g x^2	132	×	182	÷
033	×	083	STO + 7	133	RCL 4	183	f $x > 0$
034	RCL 5	084	R/S	134	+	184	GTO 3
035	f cos	085	STO 3	135	STO 5	185	2
036	RCL 3	086	STO 6	136	RCL A	186	4
037	f cos	087	g x^2	137	+	187	+
038	×	088	STO + 7	138	f sin	188	f LBL 3
039	STO 0	089	f P ←→ S	139	RCL 6	189	g → H.MS
040	RCL 1	090	RCL 7	140	×	190	R/S
041	f cos	091	h RC I	141	RCL 0	191	RCL 3
042	×	092	×	142	×	192	RCL 8
043	RCL 1	093	STO 0	143	f P ←→ S	193	÷
044	f sin	094	1	144	STO + 1	194	g $\sin^{-1}$
045	RCL 3	095	f LBL 2	145	f P ←→ S	195	g → H.MS
046	f sin	096	STO 1	146	RCL 5	196	R/S
047	×	097	3	147	RCL B	197	f P ←→ S
048	–	098	h y^x	148	+	198	RCL 0
049	g → P	099	2	149	f sin	199	R/S
050	STO D	100	×	150	RCL D	200	5

201 STO + 7	207 ×	213 RCL 6	219 f $\sqrt{x}$
202 f P ←→ S	208 RCL 2	214 ×	220 ÷
203 RCL 8	209 RCL 5	215 +	221 g $\cos^{-1}$
204 R/S	210 ×	216 RCL 8	222 f P ←→ S
205 RCL 1	211 +	217 ÷	223 h RTN
206 RCL 4	212 RCL 3	218 RCL 7	

2. Store elements of parabolic orbit:
   - ϵ, STO 1 (referred to same epoch as elements—ϵ_{1950}= 23.445 788)
   - q, STO 2
   - i, STO 3
   - ω, STO 4
   - Ω, STO 5
   - t, STO 7 (days before perihelion, as starting date for ephemeris; before perihelion, t is negative)

   Press A

3. Enter X, press B; Y, press R/S; Z, press R/S — Geocentric equatorial rectangular coordinates of the Sun, referred to same epoch as elements. See Programme 23.

4. The programme first computes the Gaussian constants for the orbit and stores them for future use in constructing an ephemeris. It then iterates for v. The programme stops to display α in H.MS format.

Press R/S. The programme stops again to display δ in D.MS format.

Press R/S. The display shows r, the radius vector from the centre of the Sun in AU.

Press R/S. The display shows Δ, the distance of the comet from the centre of the Earth, in AU.

Press R/S. The programme ends by displaying the elongation of the comet from the Sun, in degrees.

Caution: The *whole* of the calculation, including the elongation, *must* be completed for each position, even if certain data are not required, to ensure that all the P ←→ S instructions are carried out.

5. For the next ephemeris position, return to Step 3.

As written, the programme gives a 5-day ephemeris, starting from the selected t. If a different interval is desired, amend line 200 accordingly.

Test: Use the programme to confirm the results for Example 1 of Chapter 8, and obtain:

$\alpha = 22^h\,34^m\,04^s$ $\delta = +37°\,40'\,06''$

$r = 1.167\,4$ $\Delta = 1.303\,5$

elongation = 59°.407 2

(22) **HP-67**

To compute geocentric positions for comets with elliptical elements.

1. Load the programme from a magnetic card. When recording the card, set the display for four decimal places.

001	f LBL A
002	1
003	RCL 3
004	+
005	1
006	RCL 3
007	−
008	÷
009	f $\sqrt{x}$
010	STO 0
011	RCL 6
012	f sin
013	RCL 4
014	f cos
015	×
016	CHS
017	STO C
018	RCL 6
019	f cos
020	RCL 4
021	f cos
022	×
023	STO 9
024	RCL 1
025	f cos
026	×
027	RCL 4
028	f sin
029	RCL 1
030	f sin
031	×
032	−
033	STO D
034	RCL 1
035	f sin
036	STO × 9
037	RCL 1
038	f cos
039	RCL 4
040	f sin
041	×
042	STO + 9
043	RCL 6
044	f cos
045	RCL C
046	g → P
047	STO 7
048	h R ↓
049	STO A
050	RCL 6
051	f sin
052	RCL 1
053	f cos
054	×
055	RCL D
056	g → P
057	STO 8
058	h R ↓
059	STO B
060	RCL 6
061	f sin
062	RCL 1
063	f sin
064	×
065	RCL 9
066	g → P
067	STO 9
068	h R ↓
069	STO C
070	h RTN
071	f LBL B
072	f P ←→ S
073	STO 1
074	STO 4
075	g x^2
076	STO 7
077	R/S
078	STO 2
079	STO 5
080	g x^2
081	STO + 7
082	R/S
083	STO 3
084	STO 6
085	g x^2
086	STO + 7
087	RCL 0
088	RCL 9
089	×
090	f P ←→ S
091	STO 1
092	f LBL 2
093	STO D
094	f sin
095	RCL 3
096	f D ←
097	×
098	RCL 1
099	+
100	STO E
101	RCL D
102	−
103	h ABS
104	EEX
105	8
106	CHS
107	g $x > y$
108	GTO 1
109	RCL E
110	GTO 2
111	f LBL 1
112	RCL E
113	2
114	÷
115	f tan
116	RCL 0
117	×
118	g $\tan^{-1}$
119	2
120	×
121	STO 1
122	1
123	RCL E
124	f cos
125	RCL 3
126	×
127	−
128	RCL 2
129	×
130	STO D
131	RCL 5
132	STO + 1
133	RCL 1
134	RCL A
135	+
136	f sin
137	RCL 7
138	×
139	RCL D
140	×
141	f P ←→ S
142	STO + 4
143	f P ←→ S
144	RCL 1
145	RCL B
146	+
147	f sin
148	RCL 8
149	×
150	RCL D
151	×
152	f P ←→ S
153	STO + 5
154	f P ←→ S
155	RCL 1
156	RCL C
157	+
158	f sin
159	RCL 9
160	×
161	RCL D
162	×
163	f P ←→ S
164	STO + 6
165	RCL 4
166	RCL 5
167	g → P
168	g x^2
169	RCL 6
170	g x^2
171	+
172	f $\sqrt{x}$
173	STO 8
174	RCL 5
175	RCL 4
176	g → P
177	h R ↓
178	1
179	5
180	÷
181	f $x > 0$
182	GTO 3
183	2
184	4
185	+
186	f LBL 3
187	g → H.MS
188	R/S
189	RCL 6
190	RCL 8
191	÷
192	g $\sin^{-1}$
193	g → H.MS
194	R/S
195	RCL D
196	R/S
197	RCL 8
198	f P ←→ S
199	R/S
200	f P ←→ S

201	5	207	RCL 5	213	+	219	g $\cos^{-1}$
202	STO + 9	208	×	214	RCL 8	220	f P ←→ S
203	RCL 1	209	+	215	÷	221	h RTN
204	RCL 4	210	RCL 3	216	RCL 7		
205	×	211	RCL 6	217	f $\sqrt{x}$		
206	RCL 2	212	×	218	÷		

---

2. Store elements of elliptical orbit:

ϵ, STO 1 (referred to same epoch as elements—ϵ_{1950}= 23.445 788)
a, STO 2
e, STO 3
i, STO 4
ω, STO 5
Ω, STO 6
f P ←→ S
n, STO 0 (if not given in data, $n = \frac{360}{P}$)
t, STO 9 (days before, or after, perihelion; before perihelion, t is negative)
f P ←→ S (n and t are stored in the secondary registers)
Press A

---

3. Enter X, press B; Y, press R/S; Z, press R/S — Geocentric equatorial rectangular coordinates of the Sun, referred to same epoch as elements. See Programme 23.

---

4. The programme first computes the Gaussian constants for the orbit and stores them for future use in constructing an ephemeris. It then calculates the mean anomaly, M, and iterates for the eccentric anomaly E. The programme stops to display α in H.MS format.

Press R/S. The programme stops again to display δ in D.MS format.

Press R/S. The display shows r, the radius vector from the centre of the Sun, in AU.

Press R/S. The display shows Δ, the distance of the comet from the centre of the Earth, in AU.

Press R/S. The programme ends by displaying the elongation of the comet from the Sun, in degrees.

Note: Unlike Programme 21, this programme does *not* have to be run to the end. If the elongation is not required, the run can be terminated after the computation of Δ.

---

5. For the next ephemeris position, return to Step 3.

As written, the programme gives a 5-day ephemeris, starting from the selected t. If a different interval is required, amend line 201 accordingly.

Test: Use the programme to confirm the results for Example 2 of Chapter 8, and obtain:

$\alpha = 16^h 51^m 23^s$ $\delta = -21° 14' 42''$
$r = 3.956\ 5$ $\Delta = 4.627\ 5$
elongation = 42°.364 7.

---

(23) **HP-67**

To compute the geocentric equatorial rectangular coordinates *X*, *Y*, *Z*, of the Sun, referred to the equator and equinox of 1950.0, between the years 1800 and 2100.

Two cards must be read:

A, for data
B, for programme

---

1. Load the data from magnetic card A:

R_0	279.696 68	B	36 000.768 92
R_1	2 415 020	I	2 234 941
R_2	1.524 2		
R_4	35 999.049 75	S_0	2 433 282.423
R_6	1.919 46	S_1	36 524.219 9
R_7	0.016 751 04	S_2	971 690
R_8	23.452 294	S_3	–29 696
R_9	36 525		

---

2. Load the programme from magnetic card B. When recording the card, set the display for 5 decimal places.

001	f LBL A	034	×	067	f cos	100	STO C
002	STO A	035	+	068	RCL D	101	f P ←→ S
003	RCL 1	036	STO + 5	069	×	102	RCL A
004	–	037	RCL 0	070	1	103	RCL 0
005	RCL 9	038	+	071	+	104	–
006	÷	039	RCL 3	072	÷	105	RCL 1
007	STO 3	040	RCL B	073	STO 5	106	÷
008	RCL 4	041	×	074	RCL 8	107	STO A
009	×	042	+	075	RCL 3	108	4
010	RCL 2	043	RCL 3	076	7	109	7
011	–	044	g x^2	077	6	110	2
012	STO 5	045	3	078	.	111	1
013	f sin	046	3	079	8	112	CHS
014	RCL 6	047	1	080	5	113	STO 9
015	RCL 3	048	1	081	÷	114	RCL 3
016	2	049	÷	082	–	115	RCL A
017	0	050	+	083	STO 3	116	1
018	9	051	STO C	084	RCL 5	117	3
019	÷	052	RCL 7	085	RCL C	118	×
020	–	053	RCL 3	086	f sin	119	–
021	×	054	2	087	×	120	STO 4
022	5	055	3	088	RCL 3	121	h RC I
023	0	056	9	089	f cos	122	RCL A
024	h $1/x$	057	2	090	×	123	6
025	RCL 3	058	3	091	STO D	124	7
026	EEX	059	÷	092	RCL 3	125	9
027	4	060	–	093	f tan	126	×
028	÷	061	STO D	094	×	127	+
029	–	062	g x^2	095	STO E	128	RCL A
030	RCL 5	063	1	096	RCL C	129	g x^2
031	2	064	h x ←→ y	097	f cos	130	2
032	×	065	–	098	RCL 5	131	2
033	f sin	066	RCL 5	099	×	132	1

133	×	156	1	179	STO ÷ 5	202	×
134	–	157	5	180	STO ÷ 6	203	RCL D
135	STO 5	158	×	181	STO ÷ 7	204	RCL 7
136	RCL 2	159	+	182	STO ÷ 8	205	×
137	RCL A	160	STO 7	183	STO ÷ 9	206	–
138	2	161	1	184	1	207	RCL C
139	0	162	0	185	STO + 4	208	RCL 5
140	7	163	8	186	STO – 7	209	×
141	×	164	5	187	STO + 9	210	–
142	–	165	8	188	RCL C	211	f – x –
143	RCL A	166	CHS	189	RCL 4	212	RCL E
144	g x^2	167	STO 8	190	×	213	RCL 9
145	9	168	RCL A	191	RCL D	214	×
146	6	169	STO × 5	192	RCL 5	215	RCL D
147	×	170	STO × 6	193	×	216	RCL 8
148	–	171	g x^2	194	+	217	×
149	STO 6	172	STO × 4	195	RCL E	218	+
150	2	173	STO × 7	196	RCL 6	219	RCL C
151	4	174	STO × 8	197	×	220	RCL 6
152	9	175	STO × 9	198	+	221	×
153	7	176	EEX	199	f – x –	222	–
154	5	177	8	200	RCL 8	223	f P ←→ S
155	RCL A	178	STO ÷ 4	201	RCL E	224	h RTN

3. Enter the Julian Date.

Press A.

The programme flashes X_{1950} at line 199, flashes Y_{1950} at line 211, and ends by displaying Z_{1950}.

The Sun's radius vector is stored in R_5.

If the rectangular coordinates of the Sun are also required to be referred to the equinox of date, note that *X*, *Y*, *Z*, for this epoch are stored in C, D and E respectively, and the desired values can be obtained by recalling these stores, in turn, when the programme has finished.

4. For the coordinates at another time, return to the start of Step 3.

Note: The mean error of the values given by this programme, between the years 1800 and 2100, is approximately 0.000 03 AU.

Test: Find *X*, *Y*, *Z* for 0^h ET 1978, July 7 (JD 2 443 696.5), referred to 1950.0.

	X_{1950}	Y_{1950}	Z_{1950}
Result:	–0.248 51	+0.904 50	+0.392 20
From the *AE*:	–0.248 455	+0.904 498	+0.392 199

(24) **HP-67**

To compute the apparent longitude, or the geometric longitude, of the Sun, and the radius vector, to ±0°.005 for the longitude, to ±0.000 05 for the radius vector.

1. Load the data from magnetic card A:

R_0 0.005 69	S_0 2 415 020
R_4 1 582.101 5	S_1 36 525
R_5 365.25	S_2 36 000.768 92
R_6 30.600 1	S_3 279.696 68
R_7 259.183 275	S_4 0.016 751 04
R_8 1 934.142 008	S_5 –0.000 041 8
R_9 1 720 994.5	S_6 358.475 83
	S_7 35 999.049 75
B 3 306	S_8 350.737 49
I 231.19	S_9 445 267.114 2

2. Load the programme from magnetic card B:

```
001 f LBL A
002 h SF 2
003 GTO B
004 g LBL a
005 h SF 2
006 g LBL b
007 h SF 0
008 f LBL B
009 RCL 4
010 g x > y
011 h SF 1
012 h x <--> y
013 ENT ↑
014 f INT
015 STO 1
016 –
017 EEX
018 2
019 ×
020 ENT ↑
021 f INT
022 STO 2
023 –
024 EEX
025 2
026 ×
027 STO 3
028 RCL 2
029 5
030 f √x
031 g x ≤ y
032 GTO 2
033 1
034 STO – 1
035 1
036 2
037 STO + 2
038 f LBL 2
039 RCL 5
040 RCL 1
041 ×
042 f INT
043 RCL 2
044 1
045 +
046 RCL 6
047 ×
048 f INT
049 +
050 RCL 3
051 +
052 h F? 1
053 GTO 9
054 RCL 1
055 EEX
056 2
057 ÷
058 f INT
059 STO 1
060 –
061 2
062 +
063 RCL 1
064 4
065 ÷
066 f INT
067 +
068 f LBL 9
069 h CF 1
070 RCL 9
071 +
072 h F? 2
073 R/S
074 GTO C
075 g LBL c
076 h SF 0
077 f LBL C
078 f P <--> S
079 RCL 0
080 –
081 RCL 1
082 ÷
083 STO A
084 RCL 5
085 ×
086 RCL 4
087 +
088 RCL A
089 g x²
090 8
091 EEX
092 6
093 ÷
094 –
095 STO C
096 RCL A
097 RCL 7
098 ×
099 RCL 6
100 +
101 RCL A
102 g x²
103 6
104 6
105 6
106 7
107 ÷
108 –
109 STO D
110 f sin
111 RCL D
112 f cos
113 RCL C
114 –
115 g → P
116 h R ↓
117 STO E
118 2
119 ÷
120 f tan
121 1
122 RCL C
123 +
124 1
125 RCL C
126 –
127 ÷
128 f √x
129 ×
130 g tan⁻¹
131 2
132 ×
133 RCL D
134 –
135 RCL A
136 2
137 0
138 .
139 2
140 ×
141 h RC I
142 +
143 f sin
144 5
```

145	6	165	÷	185	h CF 0	205	RCL 8
146	2	166	+	186	DSP 3	206	RCL A
147	÷	167	RCL 3	187	f $x > 0$	207	×
148	+	168	+	188	GTO 4	208	−
149	RCL A	169	RCL A	189	3	209	RCL A
150	RCL 9	170	RCL 2	190	6	210	g x^2
151	×	171	×	191	0	211	4
152	RCL 8	172	+	192	+	212	8
153	+	173	RCL A	193	f LBL 4	213	1
154	RCL A	174	g x^2	194	R/S	214	÷
155	g x^2	175	RCL B	195	1	215	+
156	6	176	÷	196	RCL E	216	f sin
157	9	177	+	197	f cos	217	2
158	5	178	f P ←→ S	198	RCL C	218	0
159	÷	179	h F ? 0	199	×	219	9
160	−	180	f GSB 3	200	−	220	÷
161	f sin	181	1	201	DSP 5	221	−
162	5	182	f R ←	202	h RTN	222	RCL 0
163	5	183	g → P	203	f LBL 3	223	−
164	9	184	h R ↓	204	RCL 7	224	h RTN

3. Enter the instant *either* as a calendar date (ET; the format is YYYY.MMDDdd, in which dd are decimals of a day) *or* as a Julian Date. According to the style of date that has been entered, press:

	For geometric λ	For apparent λ
(a) CALENDAR DATE		
If JD is also required	A	f a
If JD is not required	B	f b
(b) JULIAN DATE	C	f c

4. The display gives (if asked by A or f a) the JD.

Press R/S. The display will then show the Sun's longitude in decimal degrees (the geometric longitude will be referred to the mean equinox of date).

If the radius vector is also required, press R/S.

Note: The programme does not work for calendar dates before March 1 of the year zero, but it will for Julian Dates.

Test: Find the geometric and apparent longitudes of the Sun, and the radius vector, at 0^h ET on 1978, December 6. Run the programme and obtain $\lambda_{Geom} =$ 253°.510, $\lambda_{App} =$ 253°.504, $R =$ 0.985 32. The 1978 *AE* gives $\lambda_{Geom} =$ 253°.511, $\lambda_{App} =$ 253°.504, $R =$ 0.985 35.

Heliocentric and geocentric positions of the inner planets, Mercury, Venus and Mars

The following three programmes have been devised to give:
(a) for the heliocentric position—the heliocentric longitude, in degrees, to 2 decimal places; the heliocentric latitude, in degrees, also to 2 decimal places; and the radius vector, in AU, to 5 decimal places;
(b) for the geocentric position—the difference between the geocentric longitudes of the planet and the Sun, in degrees, to 2 decimal places (positive if the planet is East of the Sun, negative if West); the geocentric longitude and latitude, in degrees, to 2 decimal places; and the distance from the planet to the Earth in AU, to 5 decimal places.

Two magnetic cards are read for each planet, the A card bearing data and the B card containing the programme. The three programmes are similar in format, but differ in some of the coefficients employed. The information to be recorded on the three A cards is given in full, as is the complete programme for Mercury; for Venus and Mars the programme is not given in full, but the changes necessary to amend the Mercury programme are listed in detail so that the three B cards can be easily recorded.

The longitudes are referred to the mean equinox of date, to an accuracy of approximately 0°.01. The accuracy is good over a range of about 3 000 years, and will be useful for historical research purposes. It is not rigorous enough for the construction of a modern ephemeris, but if by means of Programme 38 the geocentric coordinates are converted into Right Ascension and Declination, the values obtained are more accurate than the approximations of Chapter 9.

(25) **HP-67**

To compute the heliocentric or geocentric position of Mercury. The time argument is the Julian Ephemeris Date.

1. Load the data from magnetic card A:

R_2	0.776 935 222	S_0	2 415 020
R_3	100.002 135 9	S_1	36 525
R_4	358.475 83	S_2	0.494 941 889
R_5	35 999.049 75	S_3	415.205 752 2
R_7	0.387 098 6	S_4	102.279 38
R_8	0.016 751 04	S_5	149 472.515 3
		S_6	47.145 944
		S_7	1.185 208
		S_8	0.205 614 21
		S_9	7.002 881

2. Load the programme from magnetic card B:

```
001  f LBL A      004  f P<-->S     007  -          010  STO A
002  h SF 2       005  DSP 2        008  RCL 1      011  RCL 7
003  f LBL E      006  RCL 0        009  ÷          012  ×
```

```
013  RCL 6
014  +
015  STO D
016  RCL A
017  5
018  3
019  7
020  ÷
021  RCL 9
022  +
023  STO E
024  f GSB 1
025  RCL A
026  4
027  8
028  8
029  7
030  6
031  ÷
032  RCL 8
033  +
034  f P <--> S
035  STO 9
036  f GSB 2
037  STO 0
038  RCL D
039  –
040  STO 1
041  2
042  ×
043  f sin
044  RCL E
045  2
046  ÷
047  f tan
048  g x^2
049  ×
050  f D ←
051  STO – 0
052  RCL 1
053  f sin
054  RCL E
055  f sin
056  ×
057  g sin^-1
058  STO 1
059  f GSB 4
060  RCL 7
061  ×
062  STO 6
063  STO D
064  h F ? 2
065  GTO 5
066  f GSB 1
067  RCL 8
068  RCL A
069  2
070  3
071  9
072  2
073  3
074  ÷
075  –
076  STO 9
077  f GSB 2
078  STO A
079  STO – 0
080  f GSB 4
081  STO 9
082  RCL 1
083  f cos
084  STO × 6
085  RCL 0
086  f sin
087  RCL 6
088  ×
089  RCL 0
090  f cos
091  RCL 6
092  ×
093  RCL 9
094  +
095  g → P
096  h R ↓
097  f – x –
098  RCL A
099  +
100  f GSB 7
101  RCL 0
102  f cos
103  RCL 6
104  ×
105  2
106  ×
107  RCL 9
108  ×
109  RCL D
110  g x^2
111  +
112  RCL 9
113  g x^2
114  +
115  f √x
116  STO E
117  RCL 1
118  f sin
119  RCL D
120  ×
121  RCL E
122  ÷
123  g sin^-1
124  f – x –
125  RCL E
126  DSP 5
127  h RTN
128  f LBL 1
129  RCL A
130  RCL 3
131  ×
132  RCL 2
133  +
134  g FRAC
135  3
136  6
137  0
138  ×
139  RCL A
140  g x^2
141  3
142  3
143  1
144  7
145  ÷
146  +
147  STO B
148  RCL A
149  RCL 5
150  ×
151  RCL 4
152  +
153  STO C
154  h RTN
155  f LBL 2
156  9
157  h ST I
158  RCL C
159  f LBL 3
160  f sin
161  RCL 9
162  f D ←
163  ×
164  RCL C
165  +
166  f DSZ
167  GTO 3
168  2
169  ÷
170  f tan
171  1
172  RCL 9
173  +
174  1
175  RCL 9
176  –
177  ÷
178  f √x
179  ×
180  g tan^-1
181  2
182  ×
183  h ST I
184  RCL C
185  –
186  RCL B
187  +
188  h RTN
189  f LBL 4
190  1
191  RCL 9
192  g x^2
193  –
194  h RC I
195  f cos
196  RCL 9
197  ×
198  1
199  +
200  ÷
201  h RTN
202  f LBL 5
203  RCL 0
204  f GSB 7
205  RCL 1
206  f – x –
207  RCL D
208  DSP 5
209  h RTN
210  f LBL 6
211  3
212  6
213  0
214  +
215  h RTN
216  f LBL 7
217  1
218  f R ←
219  g → P
220  h R ↓
221  f x < 0
222  f GSB 6
223  f – x –
224  h RTN
```

3. Enter the Julian Date
For the heliocentric position, press A
For the geocentric position, press E

4. (a) Heliocentric: the programme pauses to flash the longitude in degrees, to 2 decimal places. At the next pause, the display flashes the latitude in degrees, also to 2 decimal places. The programme ends by displaying the radius vector in AU, to 5 decimal places.
(b) Geocentric: the programme pauses to flash the difference between the geocentric longitudes of Mercury and the Sun, in degrees (to 2 decimal places), positive if Mercury is East of the Sun, negative if West of the Sun. At the next two pauses the display flashes the longitude and latitude respectively, in degrees, to 2 decimal places. The programme ends by displaying the distance of Mercury from the Earth in AU, to 5 decimal places.

5. For the position at another date, return to Step 3.

Test: Find the heliocentric and geocentric positions of Mercury for 1978, November 17 at 0^h ET (JD 2 443 829.5).

Results:	Elongation	l or λ	b or β	r or Δ
(a) Heliocentric:		334.02	−6.74	0.389 03
The *AE* gives:		334.01	−6.74	0.389 04
(b) Geocentric:	+22.41	256.69	−2.62	0.999 90
The *AE* gives:		256.69	−2.62	0.999 91

Note: With regard to accuracy, the employment of more than 2 decimal places for the longitude or latitude is meaningless. However, if it is intended to convert these results by means of a later programme into RA and dec., it is in this case permissable to evaluate the geocentric longitude and latitude to 4 decimal places provided that the results from the coordinate conversion programme are rounded. In this way, the combination of the two programmes will give results for the inner planets to the same accuracy as is found in the tabulated positions and elongations listed annually in the *Handbook of the British Astronomical Association*.

Example: Make the necessary adjustments to the programme to obtain λ and β to 4 decimal places. Find the data for Mercury for 1978, November 23 at 0^h ET, then convert the geocentric longitude and latitude into RA and dec. by means of Programme 38. Compare with the accurate values published in the *AE*, the rounded values in the *HBAA*, and the approximate values found by the method of Chapter 9. (JD = 2 443 835.5)

	Programme 25	*AE*	*HBAA*	Chapter 9
Elongation	+20°.662 4		21°	
λ	261°.004 2			
β	− 2°.123 2			
Δ	0.868 00	0.868 031 3	0.868	0.867
RA	$17^h\,20^m\,12^s$	$17^h\,20^m\,10^s.93$		
	$= 17^h\,20^m.2$	$= 17^h\,20^m.2$	$17^h\,20^m.2$	$17^h\,20^m.5$
dec.	−25° 15′ 09″	−25° 15′ 10″.5		
	= −25° 15′	= −25° 15′	−25° 15′	−25° 16′

(26) **HP-67**

To compute the heliocentric or geocentric position of Venus. The time argument is the Julian Ephemeris Date.

---

1. Load the data from magnetic card A:

R_2	0.776 935 222	S_0	2 415 020
R_3	100.002 135 9	S_1	36 525
R_4	358.475 83	S_2	0.952 130 694
R_5	35 999.049 75	S_3	162.553 366 4
R_7	0.723 331 6	S_4	212.603 22
R_8	0.016 751 04	S_5	58 517.803 87
		S_6	75.779 647
		S_7	0.899 850
		S_8	0.006 820 69
		S_9	3.393 631

---

2. Load the programme from magnetic card B:

   This card is prepared as for Programme 25, with the following changes incorporated:

   Replace lines 17 to 19 with EEX
3

   (Lines 20 to 25 now come in the position lines 19 to 24)

   Replace lines 26 to 30 with CHS
2
0
9
4
6

   Replace lines 141 to 144 with 3
2
6
8

   Replace 9 by 4 in line 156.

3. The operation of the programme is identical to that of Programme 25.

---

(27) HP-67

To compute the heliocentric or geocentric position of Mars. The time argument is the Julian Ephemeris Date.

1. Load the data from magnetic card A:

R_2	0.776 935 222	S_0	2 415 020
R_3	100.002 135 9	S_1	36 525
R_4	358.475 83	S_2	0.815 937 028
R_5	35 999.049 75	S_3	53.171 376 42
R_7	1.523 688 3	S_4	319.519 13
R_8	0.016 751 04	S_5	19 139.854 75
		S_6	48.786 442
		S_7	0.770 992
		S_8	0.093 312 90
		S_9	1.850 333

2. Load the programme from magnetic card B:

This card is prepared as for Programme 25, with the following changes incorporated:

Replace lines 16 to 30 with
RCL 9
RCL A
1
4
8
1
÷
−
STO E
f GSB 1
RCL A
1
0
8
6
2

(Lines 31 to 140 now come in the position lines 32 to 141)

Replace lines 141 to 144 with
1
f D←
g x^2

Replace 9 by 6 in line 156.

3. The operation of the programme is identical to that of Programme 25.

Interpolation, from 3 ephemeris positions.

1. Enter the programme:

						Register contents:
01	STO 1	10	RCL 2	19	RCL 1	R_0 n
02	R/S	11	–	20	–	R_1 x_1
03	STO 2	12	STO 4	21	RCL 0	R_2 x_2
04	R/S	13	R/S	22	×	R_3 x_3
05	STO 3	14	STO 0	23	2	R_4 $x_1 + x_3 - 2x_2$
06	RCL 1	15	RCL 4	24	÷	
07	+	16	×	25	RCL 2	
08	RCL 2	17	RCL 3	26	+	
09	–	18	+	27	GTO 00	

2. Switch to RUN, f PRGM.

3. Enter the tabulated ephemeris positions at t_1, t_2 and t_3; the time for which the value of the variable x is required must fall between t_2 and t_3

   Enter x_1, press R/S
   Enter x_2, press R/S
   Enter x_3, press R/S

4. Enter the interpolation period, n; press R/S.

5. The display gives the value of x at time t.

6. For new case return to Step 3.

Test: *HBAA* gives the following 10-day ephemeris for Saturn:

1977, January 7	$\alpha = 9^h\,12^m.8$	$\delta = +17°\,03'$
17	$9^h\,10^m.0$	$17°\,17'$
27	$9^h\,06^m.8$	$17°\,32'$

Find the position at 0^h ET on 1977, January 20.
Enter the 3 values for α (in H.MS format):

9.12 48 g → H R/S
9.10 00 g → H R/S
9.06 48 g → H R/S

The interval after t_2 is 3 days exactly, so the interpolation period is 3 ÷ 10.
Enter:

3 ENT ↑ 10 ÷ R/S

The display gives the value of α in decimal hours.
Press f H.MS.
The value of 0^h ET on January 20 is $9^h\,09^m.1$.
Now follow the same procedure for δ:

17.03 g → H R/S
17.17 g → H R/S
17.32 g → H R/S
3 ENT ↑ 10 ÷ R/S

The display gives the value of δ in decimal degrees.
Press f H.MS.
The value for 0^h ET on January 20 is $+17°\,21'$.

(29) **HP-67**

Interpolation, from 3 ephemeris positions.

1. Load the programme from a magnetic card. As this is a short programme, try to incorporate it on a card bearing other short programmes, amending the Labels if necessary.

001	f LBL A	009	RCL 2	017	RCL 4	025	2
002	STO 1	010	−	018	×	026	÷
003	R/S	011	RCL 2	019	RCL 3	027	RCL 2
004	STO 2	012	−	020	+	028	+
005	R/S	013	STO 4	021	RCL 1	029	h RTN
006	STO 3	014	h RTN	022	−		
007	RCL 1	015	f LBL B	023	RCL 0		
008	+	016	STO 0	024	×		

2. Enter the 3 consecutive tabular values:

A, press A $(= t_1)$
B, press R/S $(= t_2)$
C, press R/S $(= t_3)$

} t_x must lie between t_2 and t_3

3. For interpolation, enter the interpolation interval n, and press B. The display gives the required value of x.

4. For new case, return to Step 2.

Test: The 1978 *AE* gives the following values for the RA of the Sun at 0^h ET:

June 23 $6^h 05^m 10^s.06$
June 24 $6^h 09^m 19^s.50$
June 25 $6^h 13^m 28^s.88$

Find the apparent RA of the Sun at $16^h 23^m 15^s.8$ on 1978, June 24.

Enter: 6.051 006, f H←, press A
6.091 950, f H←, press R/S
6.132 888, f H←, press R/S

The ephemeris is a daily one, so n will be expressed as a fraction of 24^h:

16.231 58, f H←, 24, ÷, press B

The display gives the required RA in decimal hours. Press g → H.MS.

The required RA of the Sun is $6^h 12^m 09^s.79$, which agrees exactly with the interpolation example on p 521 of the 1978 *AE*. Note, however, that this programme is not suitable for use with the Moon, where the rate of change in the value of x is pronounced; in this case, the 5-point interpolation programme (31) should be employed.

(30) **HP-25**

Interpolation, from 5 ephemeris positions. To be used in preference to Programme 28 when the rate of change in the value of x is marked (e.g., for interpolation in a lunar ephemeris).

1. Enter the programme:

						Register contents:
01	STO 0	14	RCL 0	27	–	R_0 n
02	RCL 4	15	g x^2	28	RCL 0	R_1 x_1
03	RCL 2	16	–	29	×	R_2 x_2
04	–	17	6	30	+	R_3 x_3
05	ENT ↑	18	÷	31	RCL 0	R_4 x_4
06	ENT ↑	19	×	32	×	R_5 x_5
07	2	20	+	33	2	
08	×	21	RCL 4	34	÷	
09	RCL 5	22	RCL 2	35	RCL 3	
10	–	23	+	36	+	
11	RCL 1	24	RCL 3	37	f H.MS	
12	+	25	2	38	GTO 00	
13	1	26	×			

2. Switch to RUN, f PRGM, f FIX 5.

3. Enter the tabulated ephemeris positions for t_1 to t_5; the time t for which the value of the variable x is required must fall between t_3 and t_4.

   Enter x_1 (H.MS or D.MS), g → H, STO 1
   x_2, g → H, STO 2
   x_3, g → H, STO 3
   x_4, g → H, STO 4
   x_5, g → H, STO 5

4. Enter interpolation interval, n; press R/S

5. The display gives the required value of x at time t, in H.MS or D.MS format.

6. For new case return to Step 3.

Test: Find the RA of the Moon at 6^h ET on 1978, November 18, given the following daily positions at 0^h ET:

November 16	$4^h\ 24^m\ 09^s.424$
17	$5^h\ 16^m\ 18^s.187$
18	$6^h\ 07^m\ 57^s.694$
19	$6^h\ 58^m\ 45^s.847$
20	$7^h\ 48^m\ 27^s.626$

In this case the interpolation interval is $\frac{1}{4}$ day (0.25).
The RA at the required time is found to be $6^h\ 20^m\ 45^s.3$.
As a check, the value listed in the *AE* is $6^h\ 20^m\ 45^s.296$.

(31) **HP-67**

Interpolation, from 5 ephemeris positions. To be used in preference to Programme 29 when the rate of change in the value of x is marked (e.g., for interpolation in a lunar ephemeris).

1. Load the programme from a magnetic card. As this is a short programme, try to incorporate it on a card bearing other short programmes, amending the labels if necessary.

001	f LBL A	015	h RTN	029	+	043	×
002	STO 1	016	f LBL B	030	1	044	–
003	2	017	STO 0	031	RCL 0	045	RCL 0
004	R/S	018	f LBL 3	032	g x^2	046	×
005	STO 2	019	RCL 4	033	–	047	+
006	3	020	RCL 2	034	6	048	RCL 0
007	R/S	021	–	035	÷	049	×
008	STO 3	022	ENT ↑	036	×	050	2
009	4	023	ENT ↑	037	+	051	÷
010	R/S	024	2	038	RCL 4	052	RCL 3
011	STO 4	025	×	039	RCL 2	053	+
012	5	026	RCL 5	040	+	054	h RTN
013	R/S	027	–	041	RCL 3		
014	STO 5	028	RCL 1	042	2		

2. Enter five consecutive tabular values from the ephemeris, selected so that t_x lies between t_3 and t_4:

A, press A
B, press R/S
C, press R/S
D, press R/S
E, press R/S

For entries 2 to 5 the calculator display provides a prompt (lines 003, 006, 009 and 012).

3. For interpolation, enter the interpolation interval n, and press B. The display gives the required value of x.

4. For new case, return to Step 2.

Test: To test the programme when first recorded, use the test example for Programme 30.

(32) **HP-25**

To find the Julian Date for any time on any day between 0 AD, January 1 and 2399, December 31. See separate instructions for dates before 0 AD.

1. Enter the programme:

01	ENT ↑	14	2	27	R/S	40	2
02	STO 4	15	×	28	RCL 5	41	÷
03	RCL 0	16	ENT ↑	29	+	42	f.INT
04	f $x \geqslant y$	17	f INT	30	RCL 7	43	1
05	GTO 09	18	STO 6	31	RCL 1	44	6
06	RCL 3	19	−	32	×	45	f $x \geqslant y$
07	STO 5	20	EEX	33	f INT	46	GTO 49
08	R ↓	21	2	34	+	47	−
09	R ↓	22	×	35	RCL 2	48	STO − 5
10	f INT	23	STO + 5	36	+	49	RCL 5
11	STO 7	24	RCL 6	37	STO 5		
12	−	25	1	38	RCL 4		
13	EEX	26	−	39	EEX		

2. Switch to RUN, f PRGM. f FIX 1. Enter constants: 1582.1015 STO 0, 365.25 STO 1, 1721057.5 STO 2, 10 CH5 STO 3.

3. Clear Register 5: 0, STO 5.

4. Enter date in YYYY.MMDD format (e.g., 1977, March 1 = 1977.03 01, 1978, January 0.5 = 1978.01 00 5, 1978, July 12.8 = 1978.07 12 8). Press R/S.

5. The programme stops at line 27 for data input. (The display shows the number of odd months stored in R_5.) According to the number displayed, enter the figure alongside it in the table:

0.0	No input required	3.0	90	6.0	181	9.0	273
1.0	31	4.0	120	7.0	212	10.0	304
2.0	59	5.0	151	8.0	243	11.0	334

6. Press R/S. The programme ends by displaying the required Julian Date, which might need amendment:

   1st Adjustment—If the required year is a leap year, and the date lies anywhere in January or February, deduct 1.0 from the display to obtain the correct Julian Date.

   2nd Adjustment—Add 1.0 to the display for all dates on or after 2000, January 1.

7. For next case, return to Step 3.

Test: Find the Julian Date for 1978, July 12.

Clear Register 5, enter 1978.07 12, and press R/S. The programme stops to display 6.0. From the table, enter 181 and press R/S. The programme ends by displaying the required Julian Date: 2 443 701.5.

The programme is still valid for BC dates but the operating procedure is changed. If the Julian Date for a day earlier than AD 0 is required, proceed as follows:

Enter the date in the usual format, followed by CHS.

At the first halt (line 27) the display will show a negative number. Key in 2, +; ignore the sign, and enter the appropriate number from the table. Then key RCL 5, CHS, STO 5, R ↓, R/S. If necessary, carry out 1st Adjustment.

(33) **HP-67**

To compute: (1) Julian Date, (2) day of week, (3) Δt (days) between two days, (4) date of New Moon.

---

1. Load the programme from a magnetic card:

001	f LBL 1	051	+	101	g FRAC	151	6
002	1	052	3	102	7	152	0
003	5	053	0	103	×	153	0
004	8	054	.	104	DSP 0	154	2
005	2	055	6	105	h RTN	155	6
006	.	056	0	106	f LBL C	156	7
007	1	057	0	107	f GSB 1	157	RCL 5
008	0	058	1	108	STO 4	158	×
009	1	059	×	109	R/S	159	7
010	5	060	f INT	110	f GSB 1	160	3
011	g $x > y$	061	+	111	RCL 4	161	.
012	h SF 2	062	RCL 3	112	–	162	6
013	h $x \longleftrightarrow y$	063	+	113	h ABS	163	3
014	ENT ↑	064	1	114	h RTN	164	+
015	f INT	065	7	115	f LBL D	165	f sin
016	STO 1	066	2	116	STO 4	166	.
017	–	067	0	117	f GSB 1	167	1
018	EEX	068	9	118	STO 5	168	7
019	2	069	9	119	.	169	4
020	×	070	5	120	0	170	3
021	ENT ↑	071	+	121	3	171	×
022	f INT	072	h F ? 2	122	3	172	STO + 7
023	STO 2	073	h RTN	123	8	173	1
024	–	074	RCL 1	124	6	174	3
025	EEX	075	EEX	125	3	175	.
026	2	076	2	126	1	176	0
027	×	077	÷	127	9	177	6
028	STO 3	078	f INT	128	2	178	4
029	RCL 2	079	STO 0	129	2	179	9
030	5	080	–	130	STO 6	180	9
031	f $\sqrt{x}$	081	2	131	×	181	2
032	g $x \leq y$	082	+	132	.	182	4
033	GTO 2	083	RCL 0	133	6	183	5
034	1	084	4	134	7	184	RCL 5
035	STO – 1	085	÷	135	0	185	×
036	1	086	f INT	136	9	186	2
037	2	087	+	137	4	187	7
038	STO + 2	088	h RTN	138	+	188	1
039	f LBL 2	089	f LBL A	139	g FRAC	189	.
040	3	090	f GSB 1	140	1	190	5
041	6	091	.	141	h $x \longleftrightarrow y$	191	+
042	5	092	5	142	–	192	STO 8
043	.	093	–	143	RCL 6	193	f sin
044	2	094	h RTN	144	÷	194	.
045	5	095	f LBL B	145	STO 7	195	4
046	RCL 1	096	f GSB 1	146	STO + 5	196	0
047	×	097	6	147	.	197	8
048	f INT	098	–	148	9	198	9
049	RCL 2	099	7	149	8	199	×
050	1	100	÷	150	5	200	STO – 7

201	RCL 8	207	1	213	.	219	RCL 4
202	2	208	6	214	5	220	+
203	×	209	1	215	−	221	DSP 4
204	f sin	210	×	216	EEX	222	h RTN
205	.	211	STO + 7	217	4		
206	0	212	RCL 7	218	÷		

2. For the Julian Date:

Enter the date in YYYY.MMDDdd format

(e.g., 1978, July 12.35 = 1978.07 12 35)

Press A; the display shows the required Julian Date.

3. For the day of week:

Enter the date in YYYY.MMDD format. Press B; the display shows 0 for Sunday, 1 for Monday, 2 for Tuesday, etc.

4. For the difference, in days, between two dates:

Enter the first date in YYYY.MMDD format, and press C.

Enter the second date in the same format, and press R/S.

The display gives the required Δt.

5. For the date of New Moon:

Enter the year and month in YYYY.MM format, and press D. The display gives the date of New Moon for that month. In some cases 0 is displayed (e.g., for 1973, July the programme gives 1973, July 0, which is interpreted as 1973, June 30).

If, instead, a specific date is entered in YYYY.MMDD format, the display will show the date of the *next* New Moon, but in that case the days may exceed 30 or 31.

For example, 1980, November 18 (1980.11 18) gives 1980, November 37, which is interpreted as 1980, December 7.

Notes:

(a) None of the four sections of this programme works for dates before March 1 of the year 0.

(b) New Moon: In about 1 per cent of the cases, the date given is 1 day in error. Also, for 1582, October, the Julian Calendar is used.

Test: After recording the programme, test it against tabulated data in any issue of the *AE*.

(34) **HP-67**

To compute the Calendar Date from the Julian Date (inverse of Programme 33).

1. Load the programme from a magnetic card:

```
001 f LBL A
002 .
003 5
004 +
005 g FRAC
006 STO 0
007 h last x
008 f INT
009 STO 1
010 2
011 2
012 9
013 9
014 1
015 6
016 1
017 g x > y
018 GTO 1
019 h x ←→ y
020 1
021 8
022 6
023 7
024 2
025 1
026 6
027 .
028 2
029 5
030 −
031 3
032 6
033 5
034 2
035 4
036 .
037 2
038 5
039 ÷
040 f INT
041 STO + 1
042 4
043 ÷
044 f INT
045 STO − 1
046 1
047 STO + 1
048 f LBL 1
049 RCL 1
050 1
051 7
052 2
053 0
054 9
055 9
056 5
057 −
058 STO 2
059 1
060 2
061 2
062 .
063 1
064 −
065 3
066 6
067 5
068 .
069 2
070 5
071 ÷
072 f INT
073 STO 3
074 RCL 2
075 RCL 3
076 3
077 6
078 5
079 .
080 2
081 5
082 ×
083 f INT
084 STO 4
085 −
086 3
087 0
088 .
089 6
090 0
091 0
092 1
093 ÷
094 f INT
095 STO 5
096 RCL 2
097 RCL 4
098 −
099 RCL 5
100 3
101 0
102 .
103 6
104 0
105 0
106 1
107 ×
108 f INT
109 −
110 EEX
111 4
112 ÷
113 STO 6
114 1
115 3
116 .
117 5
118 RCL 5
119 g x ≤ y
120 GTO 2
121 1
122 2
123 −
124 f LBL 2
125 1
126 −
127 STO 7
128 EEX
129 2
130 ÷
131 STO + 6
132 5
133 f √x
134 RCL 7
135 g x > y
136 GTO 3
137 1
138 STO +3
139 f LBL 3
140 RCL 3
141 RCL 6
142 +
143 RCL 0
144 EEX
145 4
146 ÷
147 +
148 h RTN
```

2. Enter the Julian Date, and press A.
3. The display gives the Calendar Date in YYYY.MMDDdd format.
4. For next case return to Step 2.

Test: The Calendar Date for JD = 2 443 701.835 is 1978, July 12.335.

(35) **HP-25**

To convert equatorial coordinates α, δ, to ecliptic coordinates λ, β, where λ = ecliptic longitude, and β = ecliptic latitude.

1. Enter the programme:

						Register contents:
01	g→H	18	RCL 1	35	f sin	R_0 ϵ
02	1	19	f sin	36	×	R_1 α
03	5	20	×	37	+	R_2 δ
04	×	21	–	38	RCL 3	R_3 β
05	STO 1	22	g $\sin^{-1}$	39	f cos	R_4 360
06	R/S	23	STO 3	40	÷	
07	g→H	24	RCL 0	41	g $\sin^{-1}$	
08	STO 2	25	f sin	42	RCL 3	
09	f sin	26	RCL 2	43	R/S	
10	RCL 0	27	f sin	44	$x \leftrightarrow y$	
11	f cos	28	×	45	g $x \geqslant 0$	
12	×	29	RCL 0	46	GTO 00	
13	RCL 0	30	f cos	47	RCL 4	
14	f sin	31	RCL 2	48	+	
15	RCL 2	32	f cos	49	GTO 00	
16	f cos	33	×			
17	×	34	RCL 1			

2. Switch to RUN, f PRGM, f FIX 6.
Enter constants: ϵ, STO 0; 360, STO 4 (ϵ_{1950}= 23.445 788).

3. Enter α (H.MS); press R/S.
Enter δ (D.MS); press R/S.

4. The programme stops at line 43 to display β; press R/S. The programme ends by displaying λ. Both λ and β are given in decimal degrees.
Check that $\cos\lambda \cos\beta = \cos\alpha \cos\delta$:
(f cos, RCL 3, f cos, ×, RCL 1, f cos, RCL 2, f cos, ×, –)
The display should be 0 (or, at least, less than 0.000 01).

Test: $\alpha = 22^h 35^m 15^s.24$, $\delta = +2° 10' 25''.24$; find β, λ.
Result is $\beta = 10°.281 817$, $\lambda = 341°.252 535$.
Both sets of coordinates are referred to 1950.0.

To convert ecliptic coordinates λ, β, to equatorial coordinates α, δ.

1. Enter the programme:

						Register contents:
01	STO 1	18	STO 3	35	f cos	R_0 ε
02	R/S	19	RCL 0	36	÷	R_1 β
03	STO 2	20	CHS	37	g sin⁻¹	R_2 λ
04	f sin	21	f sin	38	1	R_3 δ
05	RCL 1	22	RCL 1	39	5	R_4 24
06	f cos	23	f sin	40	÷	R_5 α
07	RCL 0	24	×	41	g x ⩾ 0	
08	f sin	25	RCL 0	42	GTO 46	
09	×	26	f cos	43	RCL 4	
10	×	27	RCL 1	44	+	
11	RCL 1	28	f cos	45	STO 5	
12	f sin	29	RCL 2	46	f H.MS	
13	RCL 0	30	f sin	47	R/S	
14	f cos	31	×	48	RCL 3	
15	×	32	×	49	f H.MS	
16	+	33	+			
17	g sin⁻¹	34	RCL 3			

2. Switch to RUN, f PRGM, f FIX 6.
   Enter constants: ε, STO 0; 24, STO 4 (ϵ_{1950}= 23.445 788).

3. Enter β (in decimal degrees); press R/S
   Enter λ (in decimal degrees); press R/S

4. The programme stops at line 47 to display α; press R/S. The programme ends by displaying δ. α is given in H.MS format, and δ is given in D.MS format.
   Check that cosδ cosα = cosβ cosλ:
   (RCL 3, f cos, RCL 5, 15, ×, f cos, ×, RCL 1, f cos, RCL 2, f cos, ×, –). The display should be 0 (or, at least, less than 0.000 01).

Test: Take the reverse of the test for Programme 35:
$\beta = 10°.281\ 817$, $\lambda = 341°.252\ 535$. The programme gives: $\alpha = 22^h\ 35^m\ 15^s.23$, $\delta = +2°\ 10'\ 25''.24$.
The result differs from the input for the Programme 35 test only by $0^s.01$ in α. Both sets of coordinates are referred to 1950.0.

(37) **HP-25**

To compute the azimuth A, and altitude λ, of a star at the observer's latitude φ.

1. Enter the programme:

						Register contents:
01	g→H	18	RCL 0	35	×	R_0 φ
02	RCL 7	19	f cos	36	RCL 0	R_1 H
03	×	20	×	37	f sin	R_2 δ
04	R/S	21	RCL 2	38	RCL 3	R_3 cosδ cos$H$
05	g→H	22	f sin	39	×	R_4 z
06	RCL 7	23	RCL 0	40	−	R_5 sin z
07	×	24	f sin	41	RCL 5	R_6 Not used
08	−	25	×	42	÷	R_7 15
09	STO 1	26	+	43	g $\cos^{-1}$	
10	R/S	27	g $\cos^{-1}$	44	R/S	
11	g→H	28	STO 4	45	9	
12	STO 2	29	f sin	46	0	
13	f cos	30	STO 5	47	RCL 4	
14	RCL 1	31	RCL 0	48	−	
15	f cos	32	f cos	49	GTO 00	
16	×	33	RCL 2			
17	STO 3	34	f sin			

2. Switch to RUN, f PRGM.
   Store constants: Observer's latitude φ (in D.MS format), g→H, STO 0; 15, STO 7.

3. Enter LST (in H.MS format); press R/S.
   Enter α (H.MS); press R/S.
   Enter δ (D.MS); press R/S.

4. The programme stops at line 44 to display the required azimuth, in decimal degrees; E of the N point of horizon if star has not yet reached the meridian (α > LST), or W of the N point of horizon if the star is past the meridian (LST > α). The programme ends by displaying the altitude, also in decimal degrees. (A negative result indicates that the star is below the horizon.)
   If the zenith distance is required, RCL 4.

5. For a new case, return to Step 3.

Test: If the observer's latitude is N 52° 03′ 26″.76, find the azimuth and altitude of a star whose equatorial coordinates are: α = $2^h\,23^m\,24^s.84$, δ = −5° 18′ 13″.8 at LST = $3^h\,41^m\,00^s$.

Result: Azimuth = 157°.48 (W of N point of horizon), altitude = 30°.30. (Both sets of coordinates are referred to 1950.0.)

(38) **HP-67**

To convert equatorial coordinates α, δ, to ecliptic coordinates λ, β, and vice versa.

1. Load the programme from a magnetic card:

001	f LBL D	018	f sin	035	3	052	×
002	h CF 0	019	×	036	6	053	RCL B
003	GTO 4	020	CHS	037	0	054	f cos
004	f LBL E	021	h F ? 0	038	+	055	×
005	h SF 0	022	CHS	039	f LBL 3	056	h F ? 0
006	f H ←	023	h RC I	040	h F ? 0	057	CHS
007	1	024	f cos	041	GTO 4	058	h RC I
008	5	025	RCL A	042	1	059	f cos
009	×	026	f sin	043	5	060	RCL B
010	f LBL 4	027	×	044	÷	061	f sin
011	STO A	028	+	045	g → H.MS	062	×
012	R/S	029	RCL A	046	f LBL 4	063	+
013	h F ? 0	030	f cos	047	R/S	064	g $\sin^{-1}$
014	f H ←	031	g → P	048	RCL A	065	h F ? 0
015	STO B	032	h R ↓	049	f sin	066	h RTN
016	f tan	033	f $x > 0$	050	h RC I	067	g → H.MS
017	h RC I	034	GTO 3	051	f sin	068	h RTN

2. Store ϵ (decimal degrees) in I. (ϵ_{1950}= 23.445 788). (Use value of ϵ for current epoch if this programme is used after Programmes 25 to 27.)

3. α, δ, to λ, β.
   Enter α (H.MS format); press E.
   Enter δ (D.MS format); press R/S.
   The programme stops to display λ in decimal degrees. Press R/S; the programme ends by displaying β in decimal degrees.

4. λ, β, to α, δ.
   Enter λ in decimal degrees; press D.
   Enter β in decimal degrees; press R/S.
   The programme stops to display α in H.MS format. Press R/S; the programme ends by displaying δ in D.MS format.

Test: Convert $\alpha = 22^h\ 35^m\ 15^s.24$, $\delta = +2°\ 10'\ 25''.24$ into λ, β. Then convert λ, β, back into α, δ.

Result: $\lambda = 341°.252\ 535$, $\beta = 10°.281\ 817$.
$\alpha = 22^h\ 35^m\ 15^s.24$, $\delta = +2°\ 10'\ 25''.24$.
(Both sets of coordinates are referred to 1950.0.)

Easter Day

There are various tables available from which the date of Easter can be established according to ecclesiastical rules. The astronomer will find Tables 14.7 (Epact), 14.9 (Gregorian Dominical Letter) and 14.10 (Gregorian Paschal table), in the *Explanatory Supplement to the AE*, most useful for this purpose. The date of Easter Day for any year after AD 1582 can be found easily and quickly. Further, if the year lies between AD 1961 and 2000, Table 14.11 gives the date directly.

Once the date of Easter has been established, other dates of religious significance can be found:

Septuagesima	–63 days	Rogation Sunday	+35 days
Sexagesima	–56 days	Ascension Day	+39 days
Quinquagesima	–49 days	Whit Sunday	+49 days
Palm Sunday	– 7 days	Trinity Sunday	+56 days

So far as the date of any current Easter is concerned, there is no real need to carry out the rather involved computation oneself, because the tables give the information quickly and reliably. That is why I have not included a method of manual computation for the HP-25 or other programmable calculators. Once such a programme has been recorded on a magnetic card, however, the situation becomes different and astronomers engaged in historical research might find this most useful. Jean Meeus has devised just such a programme, which will compute the date of Easter in the absence of any tables (but not before AD 1583).

(39) **HP-67**

To compute the date of Easter.

(The display will show 'Error' if a non-integral year, or one before AD 1583, is entered.)

---

1. Load the programme from a magnetic card:

001	f LBL A	019	CHS	037	÷	055	g FRAC
002	STO 0	020	f $\sqrt{x}$	038	STO 0	056	4
003	1	021	h RTN	039	f INT	057	×
004	5	022	f LBL 2	040	STO 2	058	f RND
005	8	023	DSP 0	041	RCL 0	059	STO 4
006	3	024	RCL 0	042	g FRAC	060	RCL 2
007	g $x \leq y$	025	1	043	EEX	061	8
008	GTO 1	026	9	044	2	062	+
009	2	027	÷	045	×	063	2
010	CHS	028	g FRAC	046	f RND	064	5
011	f $\sqrt{x}$	029	1	047	STO 0	065	÷
012	h RTN	030	9	048	RCL 2	066	f INT
013	f LBL 1	031	×	049	4	067	STO 5
014	RCL 0	032	f RND	050	÷	068	RCL 2
015	g FRAC	033	STO 1	051	STO 9	069	RCL 5
016	f $x = 0$	034	RCL 0	052	f INT	070	–
017	GTO 2	035	EEX	053	STO 3	071	1
018	2	036	2	054	RCL 9	072	+

```
073 3
074 ÷
075 f INT
076 STO 6
077 RCL 1
078 1
079 9
080 ×
081 RCL 2
082 +
083 RCL 3
084 –
085 RCL 6
086 –
087 1
088 5
089 +
090 3
091 0
092 ÷
093 g FRAC
094 3
095 0
096 ×
097 f RND
098 STO 2
099 RCL 0
100 4
101 ÷
102 STO 9
103 f INT
104 STO 7
105 RCL 9
106 g FRAC
107 4
108 ×
109 f RND
110 STO 8
111 3
112 2
113 RCL 4
114 RCL 7
115 +
116 2
117 ×
118 +
119 RCL 2
120 –
121 RCL 8
122 –
123 7
124 ÷
125 g FRAC
126 7
127 ×
128 f RND
129 STO 9
130 2
131 2
132 ×
133 RCL 1
134 +
135 RCL 2
136 1
137 1
138 ×
139 +
140 4
141 5
142 1
143 ÷
144 f INT
145 STO 1
146 7
147 ×
148 CHS
149 1
150 1
151 4
152 +
153 RCL 2
154 +
155 RCL 9
156 +
157 3
158 1
159 ÷
160 STO 0
161 f INT
162 1
163 0
164 ÷
165 RCL 0
166 g FRAC
167 3
168 1
169 ×
170 1
171 +
172 +
173 DSP 1
174 h RTN
```

2. Enter integral year for which the date of Easter is required; press A. (Non-integral years, and years before 1583, will give an 'Error' indication.)

3. The display gives the date of Easter in DD.M format, e.g., 31.3 = 31st March.

4. For new case, return to Step 2.

Test: Enter 1978, press A, and find the date of Easter Day = 1978, March 26.

(40) **HP-25**

To compute the approximate time ($\pm$ 5 min) of the rise, transit, or setting of the Sun, a planet or a star. Not suitable for the Moon unless an error of $\pm$20 min is acceptable.

1. Enter the programme:

						Register contents:
01	g → H	17	f H.MS	33	RCL 1	R_0 ST at 0^h UT
02	STO 0	18	GTO 00	34	+	R_1 α (hr) at 0^h
03	R/S	19	1	35	RCL 4	R_2 δ (deg) at 0^h
04	g → H	20	CHS	36	RCL 6	R_3 φ (deg) (lat.)
05	STO 1	21	STO 5	37	÷	R_4 λ (deg) (long.)
06	R/S	22	RCL 3	38	+	R_5 −1 or +1
07	g → H	23	f tan	39	RCL 0	R_6 15
08	STO 2	24	RCL 2	40	−	R_7 0.997 27
09	R/S	25	f tan	41	RCL 7	
10	RCL 1	26	×	42	×	
11	RCL 4	27	CHS	43	f H.MS	
12	RCL 6	28	g $\cos^{-1}$	44	GTO 00	
13	÷	29	RCL 6	45	1	
14	+	30	÷	46	STO 5	
15	RCL 0	31	RCL 5	47	GTO 22	
16	−	32	×			

2. Switch to RUN, f PRGM.
   Store constants: Enter φ (D.MS), g → H, STO 3
   λ (D.MS), g → H, STO 4
   15, STO 6
   0.997 27, STO 7

3. Enter: ST at 0^h UT of day; press R/S
   α (H.MS) at 0^h ET; press R/S
   δ (D.MS) at 0^h ET; press R/S

4. For time of rising: press GTO 19, R/S
   transit: press GTO 10, R/S
   setting: press GTO 45, R/S

5. The results are displayed in hours and minutes. If the display is negative, press g → H, 24, +, f H.MS.
   The transit times need no adjustment. Because of refraction at the horizon, the times of rising and setting require the following approximate corrections:

	Rising	Setting
For a planet or star	-2^m	$+2^m$
For the Sun (or Moon)	-3^m	$+3^m$

6. For new case, return to Step 3.

Tests: (1) Find the approximate times of rising, transit and setting of Mercury on 1978, January 7, at a location $\lambda = 0°$, φ = N 52°. GST at 0^h UT is $7^h\,04^m\,49^s.397$, $\alpha = 17^h\,32^m\,19^s.97$, $\delta = -21°\,00'\,33''.0$, and obtain: rise = $6^h\,22^m$; transit = $10^h\,27^m$; set = $14^h\,29^m$.

(2) Find the approximate times of rising, transit and setting of Jupiter on 1978, October 13, at a location $\lambda = 0°$, φ = S 35°. GST at 0^h UT is $1^h\,24^m\,47^s.980$, $\alpha = 8^h\,34^m\,19^s.073$, $\delta = +19°\,04'\,04''.00$, and obtain: rise = $2^h\,03^m$; transit = $7^h\,09^m$; set = $12^h\,13^m$.

(3) Find the approximate times of transit and sunset (upper limb) on 1978, August 3, at a location $\lambda = 0°$, $\varphi =$ N 40°. GST at 0^h UT is $20^h 44^m 52^s.750$, $\alpha = 8^h 51^m 03^s.62$, $\delta = +17° 39' 20''.6$, and obtain Sun's transit $= 12^h 06^m$; sunset $= 19^h 12^m$.

---

(41) **HP-67**

To compute the time of rise, transit or set of a planet.

---

1. Load the programme from a magnetic card:

```
001 f LBL A
002 f H ←
003 STO 0
004 3
005 6
006 0
007 .
008 9
009 8
010 5
011 6
012 4
013 7
014 STO E
015 R/S
016 f H ←
017 STO 1
018 R/S
019 f H ←
020 STO 2
021 R/S
022 f H ←
023 STO 3
024 1
025 5
026 STO × 0
027 STO × 1
028 STO × 2
029 STO × 3
030 R/S
031 f H ←
032 f P ←→ S
033 STO 1
034 R/S
035 f H ←
036 STO 2
037 R/S
038 f H ←
039 STO 3
040 f P ←→ S
041 h RTN
042 f LBL C
043 RCL 2
044 RCL A
045 +
046 RCL 0
047 –
048 3
049 6
050 0
051 ÷
052 3
053 +
054 g FRAC
055 STO 4
056 f LBL 1
057 f GSB 2
058 RCL A
059 +
060 RCL 0
061 –
062 RCL E
063 RCL 4
064 ×
065 –
066 f sin
067 g sin⁻¹
068 STO 5
069 h ABS
070 EEX
071 6
072 CHS
073 g x > y
074 GTO 3
075 RCL 5
076 3
077 6
078 0
079 ÷
080 STO + 4
081 GTO 1
082 f LBL 3
083 RCL 4
084 2
085 4
086 ×
087 g → H.MS
088 h RTN
089 f LBL B
090 1
091 CHS
092 h ST I
093 GTO 4
094 f LBL D
095 1
096 h ST I
097 f LBL 4
098 f P ←→ S
099 RCL 2
100 f P ←→ S
101 f tan
102 RCL B
103 f tan
104 ×
105 CHS
106 g cos⁻¹
107 h RC I
108 ×
109 RCL 2
110 +
111 RCL A
112 +
113 RCL 0
114 –
115 RCL E
116 ÷
117 3
118 +
119 g FRAC
120 STO 4
121 f LBL 5
122 f GSB 2
123 STO 5
124 RCL 4
125 f P ←→ S
126 STO 4
127 f GSB 2
128 f P ←→ S
129 STO 6
130 RCL E
131 RCL 4
132 ×
133 RCL 5
134 –
135 RCL A
136 –
137 RCL 0
138 +
139 STO 7
140 f cos
141 RCL B
142 f cos
143 ×
144 RCL 6
145 f cos
146 ×
147 RCL B
148 f sin
149 RCL 6
150 f sin
151 ×
152 +
153 g sin⁻¹
154 .
155 5
156 6
157 6
158 7
159 +
160 RCL 6
161 f cos
162 ÷
163 RCL B
164 f cos
```

165	÷	179	STO + 4	193	0	207	RCL 3
166	RCL 7	180	GTO 5	194	0	208	+
167	f sin	181	f LBL 6	195	÷	209	RCL 1
168	÷	182	RCL 4	196	RCL 4	210	–
169	RCL E	183	2	197	+	211	RCL 9
170	÷	184	4	198	STO 9	212	×
171	STO 8	185	×	199	RCL 3	213	2
172	h ABS	186	g → H.MS	200	RCL 1	214	÷
173	EEX	187	h RTN	201	+	215	RCL 2
174	7	188	f LBL 2	202	RCL 2	216	+
175	CHS	189	RCL C	203	–	217	h RTN
176	g $x > y$	190	8	204	RCL 2		
177	GTO 6	191	6	205	–		
178	RCL 8	192	4	206	×		

2. Store the constants:
   λ (longitude in decimal degrees, positive if W of Greenwich, negative if E), STO A
   φ (latitude in decimal degrees, negative if S), STO B
   ΔT (reduction from UT to ET, in seconds), STO C

3. Enter GST at 0^h UT (in H.MS format) for required day; press A.
   Enter α on day –1 at 0^h ET (in H.MS format); press R/S
   α on required day; press R/S
   α on day +1; press R/S
   δ on day –1 at 0^h ET (in D.MS format); press R/S
   δ on required day; press R/S
   δ on day +1; press R/S

4. To obtain the time of rise (UT, in H.MS format), press B.
   To obtain the time of transit, press C.
   To obtain the time of set, press D.

Note: The declinations are not needed if only the time of transit is required.
The times of rise and set are based on the value $h = -0° 34'$ (i.e., the zenith distance, allowing for refraction at the horizon, is $z = 90° 34'$).

Test: GST at 0^h UT on 1977, February 15 is $9^h 39^m 32^s.704$, and $\Delta T = 47^s.6$. Find the time at which Mercury rises, and the time of transit on that date for an observer on the Greenwich meridian, $\lambda = 0°$, at a latitude of $+51°.5$, given the following positions for Mercury at 0^h ET:

1977, February 14	$\alpha = 20^h 27^m 32^s.44$	$\delta = -20° 22' 25''.6$
15	$\alpha = 20^h 33^m 38^s.55$	$\delta = -20° 07' 09''.4$
16	$\alpha = 20^h 39^m 47^s.31$	$\delta = -19° 50' 35''.9$

The programme gives:
Rise at $6^h 39^m 34^s.5$ (The chart in the *HBAA* gives $6^h 40^m$ approximately.)
Transit at $10^h 55^m 05^s.9$ (The *AE* gives $10^h 55^m 06^s$).

(42) **HP-25**

To compute the approximate time, in minutes, that Mercury or Venus rises before the Sun, or sets after it, at the observer's latitude.

---

1. Enter the programme:

01	RCL 3	17	RCL 0	33	RCL 5
02	f sin	18	×	34	g $x = 0$
03	RCL 0	19	CHS	35	GTO 40
04	×	20	RCL 7	36	CLX
05	CHS	21	−	37	STO 5
06	RCL 6	22	RCL 1	38	R ↓
07	−	23	÷	39	GTO 42
08	RCL 1	24	RCL 4	40	R ↓
09	÷	25	f cos	41	CHS
10	RCL 3	26	÷	42	6
11	f cos	27	g $\cos^{-1}$	43	0
12	÷	28	RCL 3	44	×
13	g $\cos^{-1}$	29	−	45	+
14	STO 3	30	4	46	GTO 00
15	RCL 4	31	×		
16	f sin	32	RCL 2		

Register contents:
R_0 $\sin\varphi$
R_1 $\cos\varphi$
R_2 α; $\alpha - \alpha$ (planet)
R_3 δ; h
R_4 δ (planet)
R_5 Flag
R_6 0.014 54 (sin50′)
R_7 0.009 89 (sin34′)

---

2. Switch to RUN, f PRGM, f FIX 0.
Enter constants:

φ (observer's latitude) in D.MS format, g → H, f sin, STO 0
f last x, f cos, STO 1
0.014 54 STO 6
0.009 89 STO 7

---

3. Is interval before sunrise required? If yes, key 1, STO 5. (This flag will be cleared automatically at lines 36 and 37.)
Or, is interval after sunset required? If yes, key 0, STO 5 (a safety measure to ensure flag is clear).

---

4. Enter data, referred to 0^h ET on day required, in H.MS format:

α ⊙, g → H, STO 2
δ ⊙, g → H, STO 3
α (planet), g → H, STO − 2
δ (planet), g → H, STO 4

Take care when $\alpha \approx 24^h$; see Step 5.

---

5. Press R/S. The display will show the required interval, in minutes. If the result is negative, add 1 440^m.

---

6. For new case return to Step 3.

Test 1: During May and June, 1977, Mercury was a 'morning star'. Find the time, in minutes, at which Mercury rises above the E horizon before the Sun, at latitude N 51° 20′ 56″.3 on 1977, June 2, given the following data for 0^h ET:

α ⊙ $= 4^h\,39^m\,12^s.35$ δ ⊙ $= +22°\,08'\,28''.7$
α (Mercury) $= 3^h\,03^m\,06^s.15$ δ (Mercury) $= +13°\,55'\,44''.0$

Result: $+43^m$, i.e., Mercury rises 43^m before the Sun. (The chart in the *BAA Handbook* shows approximately 40^m, but this is for N 52°.)

Test 2: Given the following data for 0^h ET on 1977, February 6, find the interval in minutes between sunset and the setting of Venus, at latitude N 52°:

α ☉ $= 21^h\ 18^m\ 08^s.88$ δ ☉ $= -15°\ 42'\ 54''.0$

α (Venus) $= 0^h\ 10^m\ 08^s.30$ δ (Venus) $= +\ 2°\ 53'\ 42''.2$

Result: $-1\ 171^m$ ($= -19^h\ 31^m$), obviously impossible. Add 1440^m (see Step 5). The correct result is now displayed, 269^m ($= 4^h\ 29^m$). Alternatively, because α (Venus) is near 24^h, it could be entered as $24^h\ 10^m\ 08^s.30$, in which case the correct result would be obtained directly. (The *BAA Handbook* chart gives approximately $4^h\ 32^m$.)

---

(43) **HP-25**

To compute the time (ET) of conjunction of two planets, and their angular separation.

The HP-67 programme (44) as devised by Jean Meeus is ideal for this purpose. The same calculation *can* be carried out with the HP-25 but cannot be entirely committed to the programme memory. Thus, some manual computation must be performed.

---

1. Enter the programme:

						Register contents:
01	RCL 4	18	×	35	+	
02	RCL 2	19	+	36	g NOP	
03	−	20	RCL 4	37	RCL 6	R_0 $2(R_3)/R_2 - R_4$
04	ENT ↑	21	RCL 2	38	f $x \geqslant y$	R_1 $\alpha_1 - \alpha_2$, day −2
05	ENT ↑	22	+	39	GTO 43	R_2 $\alpha_1 - \alpha_2$, day −1
06	2	23	RCL 3	40	1	R_3 $\alpha_1 - \alpha_2$, day 0
07	×	24	2	41	STO + 7	R_4 $\alpha_1 - \alpha_2$, day +1
08	RCL 5	25	×	42	GTO 49	R_5 $\alpha_1 - \alpha_2$, day +2
09	−	26	−	43	RCL 0	R_6 10^{-9}
10	RCL 1	27	RCL 0	44	2	R_7 0
11	+	28	×	45	4	
12	1	29	+	46	×	
13	RCL 0	30	RCL 0	47	f H.MS	
14	g x^2	31	×	48	GTO 00	
15	−	32	2	49	RCL 7	
16	6	33	÷			
17	÷	34	RCL 3			

---

2. Switch to RUN, f PRGM, f FIX 4.
   Clear R_7 and enter constant:
   0, STO 7; 1, EEX, 9, CHS, STO 6.

---

3. Enter 5 data points for the RA of the first planet in R_1 to R_5, e.g., enter RA at 0^h ET for day −2 in H.MS format, g → H, STO 1, and so on.

---

4. Enter the 5 data points for the second planet, deducting each entry from the relevant memory store R_1 to R_5, e.g., enter RA at 0^h ET for day −2 in H.MS format, g → H, STO − 1, and so on.

---

5. Press: 0, STO 0, RCL 3, *2, ×, RCL 2, RCL 4, −, ÷, STO + 0, R/S.

6. When the programme has run, the display will show either the integer 1.000 0 or the time of conjunction in H.MS format. If the former, proceed as follows:

R ↓, R ↓, R ↓, return to * in Step 5 and repeat the remainder of that instruction.

After the second run, the display will show either the integer 2.0000 or the time of conjunction (R_7 is keeping a count of the programme runs). If the former, repeat the process as above (i.e., R ↓, R ↓, R ↓, and return to *). Continue in the same fashion until either the time of conjunction is displayed or the integer 8.0000 (i.e., eight programme runs, which is unlikely). There is no point in performing more than 8 runs as the difference is by now so small as to be insignificant (press R ↓, R ↓, R ↓ to see it). In this extreme case, press: RCL 0, 24, ×, f H.MS to obtain the time of conjunction.

Do not switch off at this stage if the angular separation of the two planets is also required.

7. Amend the programme. While still in the RUN mode, key GTO 35, switch to PRGM, key GTO 00, switch to RUN, f PRGM.

8. Return to Step 3 and repeat the data entry process for the dec. points of the first planet, commencing with day –2.

9. Repeat Step 4 for the declinations of the second planet.

10. Do *not* repeat Step 5.

Press RCL 0, R/S.

The amended programme will run and display the angular separation of the two planets at the time of conjunction, in decimal degrees. Press f H.MS for the separation in degrees, minutes and seconds.

A positive result indicates that the first planet lies N of the second.

Test: Find the time of conjunction and angular separation of Venus and Mars on 1977, May 13, given the same data for RA and dec. as listed in Programme 44 for the five days 11 to 15 May.

Result: The time of conjunction is $17^h\ 55^m\ 51^s$ (on 1977, May 13) (given after the 7th programme run) and the angular separation is $1°\ 17'\ 29''$ (Venus lying N of Mars). The *AE* entry gives the time of conjunction as '18 hrs, Venus 1°.3 N of Mars'.

Note: You can, if you wish, now check these results and find the RA and dec. of the two planets at the time of conjunction. Except for the last line (f H.MS) the 5-point interpolation programme for the HP-25 (Programme 30) is exactly the same as lines 01 to 35 of the conjunction programme, which we have already terminated at line 36 by changing the original NOP instruction to GTO 00 (in Step 7 above).

In other words, the 5-point interpolation programme is already in the calculator memory ready for use, except for the last line which we can enter manually.

First, enter the 5 RA points for Venus in R_1 to R_5. The interpolation period is already in R_0. Press RCL 0 to bring it into the X-register, then R/S. The display will show the RA of Venus at the time of conjunction, in decimal hours. Press f H.MS and note the RA ($0^h\ 47^m\ 18^s$).

Now enter the 5 RA points for Mars, again in R_1 to R_5. RCL 0 for the interval, press R/S. At the end of the run press f H.MS and confirm that the RA of Mars is the same as that for Venus, $0^h 47^m 18^s$. It is.

Enter the 5 declination points for Venus in R_1 to R_5, RCL 0, R/S. At the end of the run press STO 6.

Enter the 5 declination points for Mars in R_1 to R_5, RCL 0, R/S. At the end of the run press STO 7, RCL 6, $x \leftarrow\rightarrow y$, –, f H.MS, and confirm that the angular separation at conjunction is the same as that already computed, $+1° 17' 29''$. It is.

Thus we confirm that the conjunction takes place at $17^h 55^m 51^s$ on 1977, May 13, when the RA of both Venus and Mars is $0^h 47^m 18^s$, Venus lying N of Mars, and the angular separation between the two planets is $1° 17' 29''$. For the dec. of Venus, RCL 6, f H.MS. It is $+5° 05' 48''$. For the dec. of Mars, RCL 7, f H.MS. It is $+3° 48' 19''$.

---

(44) **HP-67**

(a) To compute the time (ET) of conjunction of two planets, and their angular separation.

(b) To compute the time, in minutes, that Mercury or Venus rises before the Sun, or sets after it, based on the ephemeris positions for 0^h ET.

---

1. Load the programme from a magnetic card:

001	f LBL A	029	STO 2	057	f H ←	085	DSP 4
002	f H ←	030	3	058	STO – 4	086	0
003	STO 1	031	R/S	059	5	087	STO 0
004	DSP 0	032	f H ←	060	R/S	088	8
005	2	033	STO 3	061	f H ←	089	h ST I
006	R/S	034	4	062	STO – 5	090	RCL 3
007	f H ←	035	R/S	063	f P ←→ S	091	f LBL 2
008	STO 2	036	f H ←	064	1	092	2
009	3	037	STO 4	065	R/S	093	×
010	R/S	038	5	066	f H ←	094	RCL 2
011	f H ←	039	R/S	067	STO – 1	095	RCL 4
012	STO 3	040	f H ←	068	2	096	–
013	4	041	STO 5	069	R/S	097	÷
014	R/S	042	f P ←→ S	070	f H ←	098	STO + 0
015	f H ←	043	1	071	STO – 2	099	f GSB 3
016	STO 4	044	R/S	072	3	100	f $x = 0$
017	5	045	f H ←	073	R/S	101	GTO 1
018	R/S	046	STO – 1	074	f H ←	102	f DSZ
019	f H ←	047	2	075	STO – 3	103	GTO 2
020	STO 5	048	R/S	076	4	104	f LBL 1
021	f P ←→ S	049	f H ←	077	R/S	105	RCL 0
022	1	050	STO – 2	078	f H ←	106	2
023	R/S	051	3	079	STO – 4	107	4
024	f H ←	052	R/S	080	5	108	×
025	STO 1	053	f H ←	081	R/S	109	g → H.MS
026	2	054	STO – 3	082	f H ←	110	f – x –
027	R/S	055	4	083	STO – 5	111	RCL 0
028	f H ←	056	R/S	084	f P ←→ S	112	STO A

```
113 f P ←→ S
114 RCL A
115 STO 0
116 f GSB 3
117 g → H.MS
118 h RTN
119 f LBL 3
120 RCL 4
121 RCL 2
122 –
123 ENT ↑
124 ENT ↑
125 2
126 ×
127 RCL 5
128 –
129 RCL 1
130 +
131 1
132 RCL 0
133 g x²
134 –
135 6
136 ÷
137 ×
138 +
139 RCL 4
140 RCL 2
141 +
142 RCL 3
143 2
144 ×
145 –
146 RCL 0
147 ×
148 +
149 RCL 0
150 ×
151 2
152 ÷
153 RCL 3
154 +
155 h RTN
156 f LBL E
157 h SF 2
158 f LBL D
159 f H ←
160 STO 6
161 R/S
162 f H ←
163 STO 7
164 f sin
165 RCL A
166 ×
167 CHS
168 .
169 0
170 1
171 4
172 5
173 4
174 –
175 RCL B
176 ÷
177 RCL 7
178 f cos
179 ÷
180 g cos⁻¹
181 STO 7
182 R/S
183 f H ←
184 STO – 6
185 R/S
186 f H ←
187 STO 8
188 f sin
189 RCL A
190 ×
191 CHS
192 .
193 0
194 0
195 9
196 8
197 9
198 –
199 RCL B
200 ÷
201 RCL 8
202 f cos
203 ÷
204 g cos⁻¹
205 RCL 7
206 –
207 4
208 ×
209 RCL 6
210 h F? 2
211 CHS
212 6
213 0
214 ×
215 +
216 h RTN
```

2. (a) For the time of conjunction of two planets:

Enter 5 positions of both planets at 0^h ET in the following order:

First planet α day –2, press A
α day –1, press R/S
α day 0, R/S
α day +1, R/S
α day +2, R/S
δ day –2, R/S
δ day –1, R/S
δ day 0, R/S
δ day +1, R/S
δ day +2, R/S

Second planet α day –2, R/S
α day –1, R/S
α day 0, R/S
α day +1, R/S
α day +2, R/S
δ day –2, R/S
δ day –1, R/S
δ day 0, R/S
δ day +1, R/S
δ day +2, R/S

Enter the coordinates in H.MS and D.MS format. At each data-entry point a cue number, from 1 to 5, is displayed so that you know exactly where you are in the input stage. When the programme has run the display first shows the correction to the central time, in H.MS format, to obtain the instant of conjunction in α, flashed at line 110, and the programme then continues, finally displaying $\Delta\delta$, positive if the first planet lies N of the second one.

---

3. (b) For the time, in minutes, that Mercury or Venus rises before the Sun, or sets after it:

First, store sinφ in A, cosφ in B.

Enter $\alpha \odot$ (H.MS format), press D (for rise) or E (for set)
$\delta \odot$ (D.MS format), press R/S
α (planet) (H.MS), R/S
δ (planet) (D.MS), R/S

The display will show the required time interval, in minutes. If negative, add 1440.

Test for (a): It is noted from the ephemeris that in mid-May 1977 Venus and Mars were close together in RA, Mars 'overtaking' Venus some time between 0^h ET on May 13 and May 14. Find the exact time of conjunction, and the angular distance between the two planets at that time, given the following data:

Venus, 0^h ET May	11,	$0^h 41^m 39^s.61$	$+4° 58' 36''.3$
	12,	$0^h 43^m 38^s.19$	$5° 00' 29''.8$
	13,	$0^h 45^m 41^s.94$	$5° 03' 13''.8$
	14,	$0^h 47^m 50^s.66$	$5° 06' 46''.8$
	15,	$0^h 50^m 04^s.16$	$5° 11' 06''.9$
Mars, 0^h ET May	11,	$0^h 39^m 33^s.93$	$+2° 58' 34''.2$
	12,	$0^h 42^m 22^s.75$	$3° 16' 43''.4$
	13,	$0^h 45^m 11^s.55$	$3° 34' 49''.8$
	14,	$0^h 48^m 00^s.33$	$3° 52' 53''.1$
	15,	$0^h 50^m 49^s.09$	$4° 10' 53''.1$

The correction to the central time is flashed first, and, as the central time is 0^h ET on May 13, the time of the event will obviously be the same in ET, $17^h 55^m 51^s$. The distance between the two planets is given as $+1° 17' 29''$; the result being positive, this indicates that at conjunction Venus lay to the N of Mars (in the N hemisphere Venus was above Mars).

Test for (b): Apply Test 2 for Programme 42, the HP-25 version of this section of the programme.

The result given is $-1\,171^m$. According to the instructions, add 1440 to find the correct value, 269^m ($= 4^h 29^m$). Thus, on 1977, February 6, at a latitude of 52° N, Venus set $4^h 29^m$ after sunset.

---

(45) **HP-25**

To compute the Position Angle of the Bright Limb of the Moon. (Within a few hours of Full Moon, accuracy diminishes to ±0.5.)

1. Enter the programme:

01	g → H	15	f cos	29	f cos	Register contents:
02	R/S	16	RCL 0	30	RCL 1	R_0 δ ☉
03	g → H	17	f sin	31	f sin	R_1 α ☉ − α ☾
04	STO 0	18	×	32	×	R_2 δ ☾
05	$x \leftarrow\rightarrow y$	19	RCL 0	33	$x \leftarrow\rightarrow y$	
06	R/S	20	f cos	34	÷	
07	g → H	21	RCL 2	35	g $\tan^{-1}$	
08	−	22	f sin	36	g $x \geqslant 0$	
09	1	23	×	37	GTO 00	
10	5	24	RCL 1	38	3	
11	×	25	f cos	39	6	
12	STO 1	26	×	40	0	
13	R/S	27	−	41	+	
14	STO 2	28	RCL 0	42	GTO 00	

2. Switch to RUN, f PRGM, f FIX 1.
   There are no constants to be entered.

3. Enter: α ☉ in H.MS format for time when PABL required
   R/S
   δ ☉ in D.MS format (CHS if southern dec.)
   R/S
   α ☾
   R/S
   δ ☾
   R/S

4. The display shows the PABL to one decimal place at the required time.
   For new case, return to Step 3.

Test: Find the PABL of the Moon at 0^h ET on 1978, September 23, given:

α ☉ = $11^h\,58^m\,35^s.20$ δ ☉ = + 0° 09′ 10″.7

α ☾ = $5^h\,01^m\,48^s.959$ δ ☾ = +17° 32′ 39″.85

The display gives the required PABL as 85°.5.

The *AE* gives the same value (with fraction illuminated = 0.62). But on 1978, September 18, when the fraction illuminated is 0.98, the programme gives PABL = 71°.5, while the *AE* gives 71°.4.

(46) **HP-67**

(a) To compute the angular distance between two stars;
(b) To compute the Position Angle of the Bright Limb of the Moon;
(c) Altitude of a star.

---

1. Load the programme from a magnetic card:

```
001  f LBL A
002  f H ←
003  STO 1
004  R/S
005  f H ←
006  STO 2
007  R/S
008  f H ←
009  STO − 1
010  R/S
011  f H ←
012  STO 3
013  RCL 1
014  1
015  5
016  ×
017  f cos
018  RCL 3
019  f cos
020  ×
021  RCL 2
022  f cos
023  ×
024  RCL 2
025  f sin
026  RCL 3
027  f sin
028  ×
029  +
030  g cos^-1
031  h RTN
032  f LBL B
033  f H ←
034  R/S
035  f H ←
036  STO 1
037  h x ←→ y
038  R/S
039  f H ←
040  −
041  1
042  5
043  ×
044  STO 2
045  R/S
046  f H ←
047  STO 3
048  f cos
049  RCL 1
050  f sin
051  ×
052  RCL 1
053  f cos
054  RCL 3
055  f sin
056  ×
057  RCL 2
058  f cos
059  ×
060  −
061  RCL 1
062  f cos
063  RCL 2
064  f sin
065  ×
066  h x ←→ y
067  g → P
068  h R ↓
069  f x > 0
070  h RTN
071  3
072  6
073  0
074  +
075  h RTN
076  f LBL C
077  f H ←
078  STO 0
079  f LBL 2
080  R/S
081  f H ←
082  STO 1
083  R/S
084  f H ←
085  1
086  5
087  ×
088  f cos
089  RCL 1
090  f cos
091  ×
092  RCL 0
093  f cos
094  ×
095  RCL 1
096  f sin
097  RCL 0
098  f sin
099  ×
100  +
101  g sin^-1
102  GTO 2
```

---

2. For the angular distance between two stars:

Enter α_1 (H.MS), press A
δ_1 (D.MS), press R/S
α_2 (H.MS), R/S
δ_2 (D.MS), R/S

The distance is given in decimal degrees.

---

3. For the PABL of the Moon:

Enter α ☉ (H.MS), press B
δ ☉ (D.MS), press R/S
α ☾ (H.MS), R/S
δ ☾ (D.MS), R/S

Angle P is given in decimal degrees.

---

4. For the altitude of a star:

Enter φ (D.MS), press C (φ = observer's latitude)
δ (D.MS), press R/S
H (H.MS), R/S (H = hour angle)

Angle a, the altitude of the star at the observer's latitude, is given in decimal degrees.

If φ remains the same for a second calculation, introduce only the new δ and H, and each time press R/S (*not* C).

Test for (a): Find the angular distance between two stars whose equatorial coordinates are

$\alpha_1 = 2^h\,19^m \qquad \delta_1 = 18°\,18'$

$\alpha_2 = 2^h\,03^m \qquad \delta_2 = 21°\,18'$

Result: the distance is 4°.81.

Test for (b): At 0^h on 1978, September 23 the coordinates of the Sun and Moon are

$\alpha\odot = 11^h\,58^m\,35^s.20 \qquad \delta\odot = +\,0°\,09'\,10''.7$

$\alpha☾ = 5^h\,01^m\,48^s.959 \qquad \delta☾ = +17°\,32'\,39''.85$

Run the programme and find the PABL of the Moon is 85°.5. (The *AE* gives the same value.) The difference between UT and ET can be ignored.

Test for (c): At the observer's latitude of $+52°\,03'\,26''.76$ find the altitude of a star whose declination is $-5°\,18'\,13''.8$ and whose hour angle is $-1^h\,17^m\,35^s.16$.

Result: $a = 30°.30$.

---

(47) **HP-67**

Calculation of the illuminated fraction of the Moon's disc.

---

1. Load the programme from a magnetic card:

001	f LBL A	026	2	051	f INT	076	2
002	1	027	×	052	RCL 2	077	+
003	5	028	STO 3	053	1	078	RCL 0
004	8	029	RCL 2	054	+	079	4
005	2	030	5	055	3	080	÷
006	.	031	f $\sqrt{x}$	056	0	081	f INT
007	1	032	g $x \leq y$	057	.	082	+
008	0	033	GTO 2	058	6	083	f LBL 9
009	1	034	1	059	0	084	6
010	5	035	STO – 1	060	0	085	9
011	g $x > y$	036	1	061	1	086	4
012	h SF 2	037	2	062	×	087	0
013	h $x \longleftrightarrow y$	038	STO + 2	063	f INT	088	2
014	ENT ↑	039	f LBL 2	064	+	089	5
015	f INT	040	3	065	RCL 3	090	.
016	STO 1	041	6	066	+	091	5
017	–	042	5	067	h F ? 2	092	–
018	EEX	043	2	068	GTO 9	093	RCL C
019	2	044	5	069	RCL 1	094	÷
020	×	045	STO C	070	EEX	095	STO 9
021	ENT ↑	046	EEX	071	2	096	3
022	f INT	047	2	072	÷	097	5
023	STO 2	048	÷	073	f INT	098	9
024	–	049	RCL 1	074	STO 0	099	9
025	EEX	050	×	075	–	100	9

101	.	132	.	163	6	194	×
102	0	133	9	164	.	195	f sin
103	5	134	−	165	2	196	.
104	×	135	STO 7	166	9	197	6
105	1	136	4	167	×	198	6
106	.	137	4	168	+	199	×
107	5	138	5	169	RCL 7	200	+
108	2	139	2	170	2	201	RCL 8
109	4	140	6	171	×	202	f sin
110	−	141	7	172	f sin	203	2
111	STO 8	142	.	173	.	204	.
112	4	143	1	174	2	205	1
113	7	144	1	175	1	206	×
114	7	145	RCL 9	176	4	207	−
115	1	146	×	177	×	208	1
116	9	147	RCL 9	178	+	209	f R ←
117	8	148	g x^2	179	RCL 5	210	g → P
118	.	149	6	180	2	211	h R ↓
119	8	150	9	181	×	212	h ABS
120	5	151	4	182	RCL 7	213	ENT ↑
121	RCL 9	152	÷	183	−	214	f sin
122	×	153	−	184	f sin	215	7
123	RCL 9	154	9	185	1	216	÷
124	g x^2	155	.	186	.	217	+
125	1	156	2	187	2	218	f cos
126	0	157	6	188	7	219	CHS
127	9	158	2	189	4	220	1
128	÷	159	−	190	×	221	+
129	+	160	STO 5	191	+	222	2
130	6	161	RCL 7	192	RCL 5	223	÷
131	3	162	f sin	193	2	224	h RTN

2. Enter the time (ET) in YYYY.MMDDdd format. (dd is the decimal fraction of a day, so 12^h ET = 50.) Press A.

3. The display gives the illuminated fraction of the Moon's disc, accurate to ±0.01.

The programme does not work for dates before March 1 of the year zero.

Test: Find the illuminated fraction on 1978, December 21 at 0^h ET. Run the programme and obtain 0.67. (The *AE* gives 0.66 for 0^h UT).

(48) **HP-67**

Lunar eclipses.

1. Load the data from magnetic card A:

R_0 2 299 161 — S_0 1 867 216.25
R_2 29.105 356 08 — S_1 36 524.25
R_3 25.816 918 06 — S_5 13.777 4
R_4 30.670 506 46 — S_6 1 720 995
R_5 216.637 8 — S_7 122.1
R_6 138.94 — S_8 365.25
R_7 1.847 69 — S_9 30.6001
R_8 2 415 036.025 — A 29.530 588 68
B 12.368 267
I 0.717 28

2. Load the programme from magnetic card B:

001	f LBL A	039	4	077	÷	115	RCL D
002	1	040	1	078	–	116	2
003	9	041	2	079	g 10^x	117	×
004	0	042	×	080	×	118	f sin
005	0	043	–	081	h ABS	119	6
006	–	044	RCL D	082	1	120	2
007	RCL B	045	2	083	.	121	÷
008	×	046	×	084	8	122	+
009	2	047	f sin	085	2	123	RCL C
010	–	048	8	086	1	124	f sin
011	f INT	049	.	087	6	125	.
012	STO 9	050	8	088	×	126	1
013	f LBL 0	051	÷	089	CHS	127	7
014	1	052	+	090	RCL 7	128	4
015	STO + 9	053	RCL C	091	+	129	×
016	RCL 2	054	f sin	092	RCL D	130	+
017	RCL 9	055	2	093	f cos	131	RCL E
018	×	056	.	094	3	132	2
019	f P ←→ S	057	2	095	0	133	×
020	RCL 5	058	6	096	÷	134	f sin
021	f P ←→ S	059	5	097	+	135	9
022	+	060	×	098	f $x < 0$	136	7
023	STO C	061	+	099	GTO 0	137	÷
024	RCL 3	062	RCL E	100	DSP 2	138	–
025	RCL 9	063	2	101	f – x –	139	f INT
026	×	064	×	102	RCL 8	140	STO 1
027	RCL 6	065	f sin	103	RCL 9	141	RCL 0
028	+	066	.	104	RCL A	142	g $x > y$
029	STO D	067	1	105	×	143	GTO 1
030	RCL 4	068	3	106	+	144	h x ←→ y
031	RCL 9	069	×	107	RCL D	145	f P ←→ S
032	×	070	+	108	f sin	146	RCL 0
033	RCL 5	071	f sin	109	.	147	–
034	+	072	h RC I	110	4	148	RCL 1
035	STO E	073	RCL D	111	0	149	÷
036	RCL D	074	f cos	112	6	150	f INT
037	f sin	075	3	113	×	151	f P ←→ S
038	.	076	6	114	–	152	STO + 1

153	4	171	RCL 2	189	−	207	÷
154	÷	172	RCL 3	190	EEX	208	STO + 4
155	f INT	173	RCL 8	191	4	209	5
156	STO − 1	174	×	192	÷	210	f $\sqrt{x}$
157	1	175	f INT	193	STO 4	211	RCL C
158	STO + 1	176	STO D	194	RCL 5	212	g $x > y$
159	f LBL 1	177	−	195	RCL E	213	GTO 3
160	RCL 1	178	RCL 9	196	g $x \leq y$	214	1
161	f P ←→ S	179	÷	197	GTO 2	215	STO + 3
162	RCL 6	180	f INT	198	1	216	f LBL 3
163	−	181	STO E	199	2	217	RCL 3
164	STO 2	182	RCL 2	200	−	218	RCL 4
165	RCL 7	183	RCL D	201	f LBL 2	219	+
166	−	184	−	202	1	220	f P ←→ S
167	RCL 8	185	RCL E	203	−	221	DSP 4
168	÷	186	RCL 9	204	STO C	222	R/S
169	f INT	187	×	205	EEX	223	GTO 0
170	STO 3	188	f INT	206	2		

3. Enter the year (YYYY). Press A.

The programme stops to flash (line 101) the magnitude of the first umbral eclipse, and ends by displaying the date in YYYY.MMDD format.

4. For the next umbral eclipse, press R/S.

Test: Find the date and magnitude of the first umbral eclipse for 1978. Run the programme and obtain:

Magnitude = 1.45   Date = 1978, March 24 (1978.03 24).

The *AE* gives the magnitude as 1.457.

Note: The first part of the programme may, on occasion, run for some time, as the conditions at each successive Full Moon are tested. Penumbral eclipses are not given.

(49) **HP-67**

Positions of the Galilean satellites (I–IV) of Jupiter.

1. Load the data from magnetic card A:

R_0 365.25	S_0 203.405 863
R_4 30.600 1	S_1 101.291 632 3
R_5 694 025.5	S_2 50.234 516 87
R_6 0.083 085 3	S_3 21.487 980 21
R_7 225.328	S_4 84.550 61
R_8 0.902 517 9	S_5 41.501 55
R_9 5.537 2	S_6 109.977 02
E 221.647	S_7 176.358 64
	S_8 0.985 600 3
	S_9 358.476

2. Load the programme from magnetic card B:

```
001 f LBL A
002 ENT ↑
003 f INT
004 STO 1
005 –
006 EEX
007 2
008 ×
009 ENT ↑
010 f INT
011 STO 2
012 –
013 EEX
014 2
015 ×
016 STO 3
017 RCL 2
018 5
019 f √x
020 g x ≤ y
021 GTO 2
022 1
023 STO – 1
024 1
025 2
026 STO + 2
027 f LBL 2
028 RCL 0
029 RCL 1
030 ×
031 f INT
032 RCL 2
033 1
034 +
035 RCL 4
036 ×
037 f INT
038 +
039 RCL 3
040 +
041 RCL 1
042 EEX
043 2
044 ÷
045 f INT
046 STO 1
047 –
048 2
049 +
050 RCL 1
051 4
052 ÷
053 f INT
054 +
055 RCL 5
056 –
057 h ST I
058 f P ←→ S
059 RCL 0
060 ×
061 RCL 4
062 +
063 STO A
064 h RC I
065 RCL 1
066 ×
067 RCL 5
068 +
069 STO B
070 h RC I
071 RCL 2
072 ×
073 RCL 6
074 +
075 STO C
076 h RC I
077 RCL 3
078 ×
079 RCL 7
080 +
081 STO D
082 h RC I
083 RCL 8
084 ×
085 RCL 9
086 f P ←→ S
087 +
088 STO 1
089 h RC I
090 RCL 6
091 ×
092 RCL 7
093 +
094 h x ←→ I
095 RCL 8
096 ×
097 RCL E
098 +
099 STO 2
100 RCL 1
101 2
102 ×
103 f sin
104 5
105 0
106 ÷
107 RCL 1
108 f sin
109 .
110 5
111 2
112 1
113 ÷
114 +
115 STO + 2
116 h RC I
117 2
118 ×
119 f sin
120 .
121 1
122 6
123 7
124 3
125 ×
126 h RC I
127 f sin
128 RCL 9
129 ×
130 +
131 STO 3
132 STO – 2
133 2
134 8
135 .
136 0
137 7
138 RCL 2
139 f cos
140 1
141 0
142 .
143 4
144 0
145 6
146 ×
147 –
148 f √x
149 STO 1
150 RCL 2
151 f sin
152 h x ←→ y
153 ÷
154 g sin^-1
155 h ST I
156 1
157 7
158 3
159 STO ÷ 1
160 RCL A
161 f GSB 4
162 RCL 0
163 f GSB 6
164 5
165 .
166 9
167 0
168 6
169 ×
170 f – x –
171 RCL B
172 f GSB 4
173 RCL 1
174 f GSB 6
175 9
176 .
177 3
178 9
179 7
180 ×
181 f – x –
182 RCL C
183 f GSB 4
184 RCL 2
185 f GSB 6
186 1
187 4
188 .
189 9
190 8
191 9
192 ×
193 f – x –
194 RCL D
195 f GSB 4
196 RCL 3
197 f GSB 6
198 2
199 6
200 .
201 3
202 6
203 4
204 ×
205 h RTN
206 f LBL 4
207 h RC I
208 +
209 RCL 3
210 –
211 RCL 1
212 f P ←→ S
213 h RTN
214 f LBL 6
215 f P ←→ S
216 ×
217 –
218 f sin
219 h RTN
```

3. Enter the date in YYYY.MMDDdd format. (dd is the decimal fraction of a day, so 12^h UT = 50.) Press A.

4. The display flashes, in turn, the X coordinate of Satellites I–III, and ends by displaying the X coordinate of IV. The coordinates are given with respect to the centre of the disc of Jupiter, and are in units of the planet's equatorial radius (*not* the diameter). Positive values indicate that the satellite(s) are W of Jupiter, negative values E.

Note: Valid only for dates in the Gregorian Calendar. The results are good, but not rigorous. (See *J. Br. Astron. Assoc.* **72,** 80 (1962).)

Test: Find the positions of Satellites I–IV at 0^h UT on 1978, September 13. Run the programme and obtain:

I = 5.74 r II = –9.20 r III = 14.73 r IV = 23.51 r

If the equatorial diameter is known from the ephemeris, these values can be converted into seconds of arc by multiplying by $\frac{1}{2}d$.

(50) **HP-25**

Transits and elongations of Polaris, to $\pm 2\frac{1}{2}$ minutes.

1. Enter the programme:

01	RCL 0	13	GTO 16	25	2	Register contents:
02	RCL 1	14	$x \leftarrow\rightarrow y$	26	4	R_0 Day Number; LTT (UT)
03	+	15	–	27	+	R_1 Days to end of last
04	RCL 2	16	RCL 6	28	ENT ↑	month
05	×	17	+	29	ENT ↑	R_2 ST gain over MT
06	RCL 3	18	RCL 4	30	RCL 5	R_3 ST 0^h UT January 0
07	+	19	g → H	31	×	R_4 α (Polaris) (H.MS)
08	2	20	$x \leftarrow\rightarrow y$	32	–	R_5 MT loss over ST
09	4	21	–	33	STO 0	R_6 φ (hours)
10	$x \leftarrow\rightarrow y$	22	g $x < 0$	34	f H.MS	R_7 $\frac{1}{4}$ sidereal day
11	f $x \geqslant y$	23	GTO 25	35	GTO 00	
12	GTO 14	24	GTO 28			

2. Switch to RUN, f PRGM, f FIX 4.

3. Enter constants:
   0.065 709 61, STO 2 (Daily rate of gain of ST over MT)
   GST at 0^h UT on January 0 of year, STO 3 (Decimal hours)
   Mean RA of Polaris on June 30 of year, STO 4 (H.MS format)
   0.002 730 434, STO 5 (Hourly rate of MT loss over ST)
   φ, STO 6 (In decimal hours, + if E, – if W of Greenwich)
   5.983 617 398, STO 7 ($\frac{1}{4}$ sidereal day, in decimal hours of MT)

4. Enter day of month, STO 0.
   Enter, from table, number of days to the end of last month, STO 1.

5. For time of local upper transit of Polaris on that day, press R/S.
   The display gives the required UT of transit, in H.MS format.

6.	For next W elongation:	RCL 0, RCL 7, +, f H.MS
	For last E elongation:	RCL 0, RCL 7, −, f H.MS
	For next lower transit:	RCL 0, RCL 7, 2, ×, +, f H.MS
	For last lower transit:	RCL 0, RCL 7, 2, ×, −, f H.MS
	For next E elongation:	RCL 0, RCL 7, 3, ×, +, f H.MS
	For last W elongation:	RCL 0, RCL 7, 3, ×, −, f H.MS

If, in Step 6 operation, the display shows a negative value, key: g → H, 24, +, f H.MS. The resulting time refers to the previous day; if the display exceeds 24^h, key: g → H, 24, −, f H.MS. The resulting time refers to the next following day.

Previous month	Dec	Jan	Feb	Mr	Ap	May	Jun	Jul	Aug	Sep	Oct	Nov
Ordinary year	0	31	59	90	120	151	181	212	243	273	304	334
Leap year	0	31	60	91	121	152	182	213	244	274	305	335

Note: The programme gives the UT at which the events occur on the observer's own meridian, but the results are not rigorous. However, for the purpose of setting up or checking the alignment of an equatorial head of a telescope, according to the recommended methods outlined in Sidgwick's *Amateur Astronomer's Handbook*, Chap. 16, or *Norton's Star Atlas*, 16th Edn., p 114, the adjustments may be safely carried out within ±10 minutes of transit and elongation, and the accuracy achieved by the programme is good enough to put the displayed time near the centre of this time span.

Test: Find the approximate local upper transit time of Polaris for 1976, October 23 at a location of 0° 35′ 54″.4 E of Greenwich, given ST at 0^h UT on January 0 for that year was 6.586 474 722, and the mean place for Polaris at June 30 was $2^h\ 08^m\ 43^s.2$. The result is $0^h\ 00^m\ 12^s$ UT. The next W elongation is at $5^h\ 59^m\ 13^s$; the next lower transit is at $11^h\ 58^m\ 14^s$; the next E elongation is at $17^h\ 57^m\ 15^s$. Now, if another ¼ sidereal day is added, we obtain $23^h\ 56^m\ 16^s$, i.e., there are two upper transits on 1976, October 23 at this location. (In this example, the times given by the programme are in error by $-2^m\ 19^s$, within the claimed accuracy limit.)

Chebyshev Coefficients

It was mentioned in Chapter 6 that interest is beginning to be shown by national almanac offices in providing and using data in a form other than the traditional tabulated ephemerides in large annual publications. Where users are able to provide simple computing facilities, the new tables have the advantage that the cost of publishing can be kept low; also, data in the form of Chebyshev coefficients provide a means of computing accurate ephemerides for, say, the positions of the satellites of the giant planets, which would take up a disproportionate amount of space in a normal tabulated annual ephemeris. There are drawbacks, however, as pointed out in Chapter 6.

In this early experimental period, the versions which have already been published by the USNO and the *Bureau des Longitudes* are broadly similar in concept, but the method of computation is different. In the one, the coefficients are used in the order a_0 to a_n, while the other works in the reverse order, from a_n to a_0, so it is not possible to give a single programme which is compatible with both publications. What I have done, therefore, is to give two HP-25 programmes, one suitable for each publication, and one HP-67, which is valid only for the French publication.

(51) **HP-25**

A programme for use with the *Connaissance des Temps, nouvelle série*, **employing Chebyshev coefficients.**

1. Enter the programme:

01	RCL 0	18	R/S	35	RCL 5	Register contents:
02	−	19	$x \longleftrightarrow y$	36	×	R_0 t_0
03	RCL 6	20	R ↓	37	2	R_1 *DT*
04	+	21	RCL 4	38	×	R_2 1, 2, 3, etc.
05	2	22	×	39	+	R_3 for $\cos\theta$
06	×	23	+	40	STO 4	R_4 $\cos 2\theta$
07	RCL 1	24	STO 6	41	RCL 2	R_5 $\cos 3\theta$, etc.
08	÷	25	RCL 7	42	R/S	R_6 0; frac. t; f(x)
09	1	26	g $x = 0$	43	$x \longleftrightarrow y$	R_7 n, $n-1$, etc.
10	STO 3	27	GTO 48	44	R ↓	
11	STO − 7	28	1	45	×	
12	−	29	STO + 2	46	STO + 6	
13	STO 4	30	STO − 7	47	GTO 25	
14	STO 5	31	RCL 3	48	RCL 6	
15	CLx	32	CHS	49	f FIX 4	
16	R/S	33	RCL 4			
17	RCL 2	34	STO 3			

2. Switch to RUN, f PRGM.

3. Enter constants for relevant time period from publication:

t_0, STO 0

DT, STO 1

4. f FIX 0

0, STO 6

1, STO 2

n, STO 7 (highest number for a in left-hand column)

Enter integral part of t (time for which f (x) is required)

ENT.↑

Enter decimal fraction of t (if any), STO 6, R ↓, (if t is an integer, ignore this instruction; zero has already been stored in R_6).

5. Press R/S; the programme will halt to display 0.

Key in a_0 and press R/S; the programme will stop again to display 1.

Key in a_1 and press R/S.

Continue to key in the coefficients as indicated by the cue numbers displayed by the X-register each time the programme stops.

When the last coefficient a_n has been entered, the programme will branch at the conditional test (line 26) because the content of R_7 is now zero, and the programme will end by displaying the required f (x) to 4 decimal places. If it should be required again, f (x) is stored in R_6.

6. For f (x) at another time t in the same time period covered by the table, return to Step 4; otherwise, for new case, return to Step 3.

Note: When using some tables, the value of f (x) may be given in seconds of arc; to convert to degrees, minutes and seconds, if required, key: 3600, ÷, f H.MS.

In other cases, the value displayed at the end of the programme run may be in seconds of time; further, this may be a negative value. The user will have to consider the nature of the quantities displayed and decide how he is to convert them into the desired standard units (see Test 1).

Test 1: Required to find Apparent GST at 0^h UT on 1978, October 23. The *Connaissance des Temps, nouvelle série*, gives the following table (page A2):

From 16 October to 3 November, 0^h. $DT = 18$ days

0	7 926.600 9
1	779 728.975 2
2	−0.005 4
3	0.000 1
4	0.005 4
5	0.000 3
6	−0.000 6
7	−0.000 1

The figures on the left are the subscripts for a_0 to a_n; the main column lists the Chebyshev coefficients. From page F2 note that $t_0 = 289$ (16 October) and $t = 296$ (23 October). From the table heading, $DT = 18$, while the highest value for $n = 7$; $a_0 = 7\,926.600\,9$, and so on.

Enter the data and run the programme. The value of f(x) displayed is –165 346^s.496 6. For the most accurate conversion to hours, minutes and seconds, proceed manually as follows:

Add the number of seconds in 1 day: 86 400, +
The display is still negative (–78 946.496 6)
So, in this example, add another day, in seconds: 86 400, +
The display is now positive (7 453.503 4)
Now convert: g FRAC, STO 7, f last x, f INT, 3 600, ÷, f H.MS.
Write down the hours, minutes, and integral seconds: 2^h 04^m 13^s.
RCL 7, f FIX 3.
Write the fractional seconds: 0^s.503.
Result: 2^h 04^m 13^s.503 (The *AE* gives the same value).

Test 2: The previous test example was worked for 0^h UT. In practice this would be unusual, as this value is already tabulated in the normal ephemeris. One would be more likely to require the GST at, for example, 21^h 35^m 30^s UT, and this is where the programme really comes into its own.

t is entered according to the instructions; in this case it would be 296.899 652 778, and we would find GST at 21^h 35^m 30^s on 1978, October 23 to be 23^h 43^m 16^s.321, which is exact.

---

(52) **HP-25**

A programme for use with the *Almanac for Computers* (USNO), employing Chebyshev coefficients.

---

1. Enter the programme:

						Register contents:
01	RCL 0	14	RCL 2	27	RCL 7	R_0 A
02	÷	15	RCL 3	28	g $x = 0$	R_1 B
03	RCL 6	16	×	29	GTO 33	R_2 $2x$
04	RCL 0	17	RCL 4	30	1	R_3
05	÷	18	–	31	STO – 7	R_4 $\left.\right\}$ b_{n+2} to b_0
06	+	19	+	32	GTO 12	R_5
07	RCL 1	20	RCL 4	33	RCL 3	R_6 t
08	+	21	STO 5	34	RCL 5	R_7 n
09	2	22	RCL 3	35	–	
10	×	23	STO 4	36	2	
11	STO 2	24	R ↓	37	÷	
12	RCL 7	25	R ↓	38	f FIX 6	
13	R/S	26	STO 3	39	GTO 00	

---

2. Switch to RUN, f PRGM.

---

3. Enter constants for relevant time period, from publication:
A, STO 0
B, STO 1

4. f FIX 0.
Enter: 0, STO 3, STO 4, STO 5.
Enter: Highest value for n, STO 7 (highest number in left-hand column).
Integral part of t (time for which f (x) is required), STO 6, f STK.
Fractional part of t (if t is an integer, ignore this instruction).

5. Press R/S; at first halt, n is displayed; enter a_n.
Press R/S; at second halt, n_{-1} is displayed; entre a $_{n-1}$.
Press R/S; continue until a_0 is entered.
Press R/S. The programme will branch at the conditional test (line 28), and will end by displaying f (x) to 6 decimal places.

6. For f (x) at another time in the same time period covered by the table, return to Step 4; otherwise return to Step 3.

Note: f (x) might be displayed in hours, seconds of time, or seconds of arc, etc. The user will have to refer to the table of coefficients to find out in what units f (x) is given and if an alternative form is required this can either be keyed manually after the programme run, or the necessary conversion steps may be incorporated in the programme after line 38, ending with a GTO 00 instruction.

Test: Required to find apparent GST at 0^h UT on 1977, September 20, correct to $\pm 0^s.01$.

Following the instructions in the USNO Circular (155), we find that it is only necessary to employ coefficients a_0 to a_{19} for this degree of accuracy (i.e., the absolute values of a_{20} to a_{33}, when summed, are less than $0^h.000\,002\,78$, so these coefficients may be ignored). The table in the circular, up to and including a_{19}, is:

Days 182 to 276 — 1977, July 1 to October 3
$A = 47.5$ — $B = -4.83157895$

Apparent GST at 0^h UT

0	43.433 548 11	7	0.000 000 21	14	−0.000 000 12
1	3.121 193 71	8	−0.000 000 22	15	−0.000 001 07
2	−0.000 01	9	0.000 000 91	16	0.000 000 06
3	0.000 000 08	10	0.000 000 32	17	−0.000 000 21
4	0.000 000 81	11	0.000 000 91	18	0.000 000 13
5	−0.000 000 53	12	−0.000 000 36	19	0.000 001 34
6	−0.000 000 15	13	0.000 000 14	(20 to 33 ignored)	

Enter the data according to the instructions at Steps 3 and 4, including $n = 19$ in R_7 and $t = 263$ (day 263 of 1977) in R_6.

Run the programme; at each halt the cue number displayed indicates the subscript of a which is to be entered, e.g., at the first halt 19 is displayed so 0.000 001 34 is entered at this point. When a_0 has been entered the programme ends by displaying the required apparent GST at 0^h, in decimal hours, 23.918 035. To convert to hours, minutes and seconds, key f H.MS and obtain $23^h\,55^m\,04^s.92$. The *AE* gives the apparent GST for 0^h UT on 1977, September 20 as $23^h\,55^m\,04^s.929$. The error is within the set limit of $\pm 0^s.01$.

The user is reminded that some of the tables of Chebyshev coefficients are prepared by the USNO for evaluation at 0^h UT or 0^h ET. Thus, unlike the coefficients tabled in the *Connaissance des Temps*, not all the USNO tables can be employed for evaluation of f (x) at a time other than 0^h. To this extent, therefore, the American tables might be considered less flexible in use than the French versions, and usually necessitate the use of a greater number of coefficients in order to achieve the same degree of accuracy. But, on the other hand, the American tables are valid over a longer period of time, and in certain applications this may be a distinct advantage.

(53) **HP-67**

A programme for use with the *Connaissance des Temps, nouvelle série*, **employing Chebyshev coefficients. This programme is not suitable for use with the USNO publication** *Almanac for Computers.*

1. Load the programme from a magnetic card:

001	f LBL A	019	h RTN	037	×	055	h ST I
002	STO A	020	f LBL 0	038	+	056	(i)
003	h CF 3	021	h pause*	039	h RTN	057	f DSZ
004	f GSB 0	022	h F? 3	040	f LBL E	058	(i)
005	STO B	023	h RTN	041	RCL A	059	RCL E
006	f GSB 0	024	GTO 0	042	−	060	×
007	f P ←→ S	025	f LBL 8	043	2	061	+
008	STO 9	026	RCL D	044	×	062	f LBL 5
009	f P ←→ S	027	CHS	045	RCL B	063	f DSZ
010	n ST I	028	RCL E	046	÷	064	GTO 6
011	f LBL 9	029	STO D	047	1	065	GTO 7
012	f GSB 0	030	RCL C	048	STO D	066	f LBL 6
013	STO (i)	031	×	049	−	067	f GSB 8
014	f DSZ	032	2	050	STO E	068	GTO 5
015	GTO 9	033	×	051	STO C	069	f LBL 7
016	f GSB 0	034	+	052	f P ←→ S	070	f GSB 8
017	STO 0	035	STO E	053	RCL 9		*
018	h π	036	(i)	054	f P ←→ S		h RTN

2. Enter t_0; press A.

Then, during the pauses, enter the following data; the information will be automatically stored owing to Flag 3:

DT

The highest value for n (restricted to the range 2 to 18)

a_0

a_1, and so on, until

a_n. After this last value has been entered, π will appear in the display.

3. For each computation in the same period covered by the table, enter *t*. Press E.

Note: If required, after line 070, enter any conversion instructions in the space marked *, i.e., if the units given by the table are in seconds of arc, enter 3 600, ÷, f H.MS, and end with the h RTN instruction; the result will then be obtained in degrees, minutes and seconds.

* Or, if you prefer, R/S.

Test: Find the radius vector of the Sun on 1978, July 2 at $17^h\,28^m\,00^s$.

The table in the Connaissance des Temps is:

From June 0 to 3 July 0^h $DT = 33$ days

0	1.015 571 31
1	0.001 406 08
2	−0.000 290 24
3	0.000 014 38
4	−0.000 007 45
5	−0.000 004 62
6	0.000 000 69
7	0.000 000 81
8	0.000 000 13

In this case, $t_0 = 151$ (June 0, table on page F2)

$DT = 33$

$n = 8$

$t = 183.727\,777\,8$ (July 2, table on page F2 = 183)

Run the programme and find the radius vector to be 1.016 687 4.

Occultations

The appendix concludes with programmes for the rigorous calculation of the occultation of a star by the Moon at any given place. The first, for a normal occultation, gives the time of the immersion or emersion, the position angle, the least distance, and the altitude of the star. The second, for a grazing occultation, gives the northern and southern limits, the time (UT), altitude of the star, and position angle.

Each of the two types of computation requires two programme cards, but the data-entry programme is common to both and is entered first. The arrangement of these three programmes is:

(54) (55)A—Common data input programme.
(54)B —Occultations.
(55)B —Grazing occultations.

It is assumed that the user will be an experienced observer with a full understanding of the principles of computation described in the *Explanatory Supplement to the AE*, Section 10, and Chauvenet's *Manual of Spherical and Practical Astronomy*, Chap. 10. Further, that the constants $\rho \sin\varphi'$ and $\rho \cos\varphi'$ will already have been determined from

$$\tan\varphi' = [0.993\,305\,4 + (0.11 \times 10^{-8}h)]\tan\varphi$$
$$\rho = 0.998\,327\,07 + 0.001\,676\,44\cos 2\varphi - 10^{-8}(352\cos 4\varphi - 15.7h) + 10^{-8}\cos 6\varphi$$

where φ' = geocentric latitude
φ = geographic latitude
ρ = geocentric distance in equatorial radii
h = height in metres above sea level

Thus, for Uccle, in Belgium, where $\varphi = +50°\,47'\,55''.0$, $\lambda = -4°\,21'\,29''.2$, and $h = 105$ m, then $\varphi' = 50°.609\,986$, $\rho = 0.998\,009\,8$, $\rho\sin\varphi' = 0.771\,306$ and $\rho\cos\varphi' = 0.633\,333$.

Similarly, for the observatory at Rainham, Kent, where $\varphi = +51°\,20'\,56''.3$, $\lambda = -0°\,35'\,54''.4$, and $h = 84$ m, then $\rho\sin\varphi' = 0.777\,335$ and $\rho\cos\varphi' = 0.625\,863$.

After the occultation programmes I have given, for the benefit of those who do not have ready access to the *Explanatory Supplement to the AE*, a brief review of the search technique for occultation predictions, followed by an example for one morning in August 1978, plus a programme for the reduction of the mean place of a star to its apparent place at any integral hour of ET. Finally, for those who prefer to work from the Besselian elements for an occultation, I include, as the last programme in this Appendix, a programme for computing these elements; they are not used in the occultation programmes given here, which are rigorous.

(54) (55)A **HP-67**

Data entry for occultation programmes 54B and 55B.

1. Load the programme from a magnetic card:

```
001  f LBL 0
002  h CF 3
003  h pause*
004  h F? 3
005  h RTN
006  GTO 0
007  f LBL A
008  f H ←
009  STO A
010  f GSB 0
011  f H ←
012  STO B
013  f GSB 0
014  f H ←
015  RCL A
016  –
017  1
018  5
019  ×
020  STO C
021  f GSB 0
022  ENT ↑
023  f GSB 0
024  3
025  6
026  0
027  0
028  ÷
029  –
030  STO 0
031  9
032  h ST I
033  f P ←→ S
034  f LBL 1
035  f GSB 0
036  f H ←
037  STO (i)
038  f DSZ
039  GTO 1
040  RCL 9
041  f GSB 4
042  RCL 6
043  STO E
044  RCL 3
045  f sin
046  h ST I
047  f P ←→ S
048  f GSB 2
049  STO 1
050  f GSB 3
051  f P ←→ S
052  STO 0
053  RCL 8
054  f GSB 4
055  RCL 5
056  STO E
057  RCL 2
058  f sin
059  h ST I
060  f P ←→ S
061  f GSB 2
062  STO 2
063  f GSB 3
064  f P ←→ S
065  STO 2
066  RCL 7
067  f GSB 4
068  RCL 4
069  STO E
070  RCL 1
071  f sin
072  h ST I
073  RCL 0
074  STO 1
075  f P ←→ S
076  f GSB 2
077  STO 3
078  f GSB 3
079  f P ←→ S
080  STO 3
081  RCL 1
082  –
083  2
084  ÷
085  STO E
086  f P ←→ S
087  RCL 3
088  RCL 1
089  –
090  2
091  ÷
092  STO D
093  RCL 3
094  STO + 1
095  RCL 2
096  STO – 1
097  STO – 1
098  2
099  STO ÷ 1
100  RCL D
101  STO 3
102  f P ←→ S
103  RCL 3
104  STO + 1
105  RCL 2
106  STO – 1
107  STO – 1
108  2
109  STO ÷ 1
110  RCL E
111  STO 3
112  .
113  2
114  6
115  2
116  5
117  2
118  CHS
119  STO 0
120  3
121  .
122  6
123  6
124  9
125  7
126  9
127  STO 4
128  f P ←→ S
129  1
130  5
131  .
132  0
133  4
134  1
135  0
136  6
137  8
138  5
139  STO 4
140  RCL 0
141  ×
142  RCL C
143  +
144  STO C
145  .
146  9
147  9
148  6
149  6
150  4
151  7
152  1
153  8
154  7
155  STO 9
156  h π
157  h RTN
158  f LBL 2
159  RCL D
160  f sin
161  RCL E
162  f cos
163  ×
164  h RC I
165  ÷
166  h RTN
167  f LBL 3
168  RCL E
169  f sin
170  RCL B
171  f cos
172  ×
173  RCL E
174  f cos
175  RCL B
176  f sin
177  ×
178  RCL D
179  f cos
180  ×
181  –
182  h RC I
183  ÷
184  h RTN
185  f LBL 4
186  RCL A
187  –
188  1
189  5
190  ×
191  STO D
192  h RTN
```

* Or, if you prefer, R/S.

2. Enter apparent α (H.MS format) of star; press A. During the pauses, enter the following data, which is read automatically:

Apparent δ (D.MS format) of star
Apparent GST at 0^h UT (H.MS)
The central hour T_2 (ET), an integer
$\Delta T =$ ET – UT, in seconds of time
α_1 of the Moon at time T_1 (H.MS)
α_2 ,, ,, ,, T_2
α_3 ,, ,, ,, T_3
δ_1 ,, ,, ,, T_1 (D.MS)
δ_2 ,, ,, ,, T_2
δ_3 ,, ,, ,, T_3
π_1 ,, ,, ,, T_1 (D.MS)
π_2 ,, ,, ,, T_2
π_3 ,, ,, ,, T_3

3. After the above entries, $\pi = 3.14$ is displayed. The necessary data are stored in the appropriate registers. Now, for an ordinary occultation go to Programme 54B, or for a grazing occultation go to Programme 55B.

(54)B **HP-67**

Occultation of a star by the Moon at a given place (rigorous calculation).

1. Having first loaded Programme 54/55A, and the quantities α ☾ to π_3 as indicated now enter:

   φ (in decimal degrees), STO 5

   λ (decimal degrees, negative if E of Greenwich), STO 7

   $\rho \sin\varphi'$, STO 8

   $\rho \cos\varphi'$, STO 9

2. Load the following programme from a magnetic card:

```
001 f LBL B
002 h SF 0
003 GTO 0
004 f LBL D
005 h CF 0
006 f LBL 0
007 0
008 STO 6
009 4
010 h ST I
011 f LBL 2
012 f GSB 3
013 f P <--> S
014 RCL 5
015 RCL 8
016 ×
017 RCL 6
018 RCL 7
019 ×
020 –
021 RCL 9
022 ÷
023 RCL 4
024 ×
025 g x²
026 1
027 h x <--> y
028 –
029 f √x
030 RCL 9
031 ÷
032 RCL 4
033 ÷
034 f P <--> S
035 h F? 0
036 CHS
037 STO + 6
038 f DSZ
039 GTO 2
040 RCL 6
041 RCL 0
042 +
043 g → H.MS
044 DSP 4
045 f – x –
046 f P <--> S
047 RCL 5
048 RCL 6
049 g → P
050 GTO 4
051 f LBL C
052 0
053 STO 6
054 4
055 h ST I
056 f LBL 1
057 f GSB 3
058 f DSZ
059 GTO 1
060 RCL 6
061 RCL 0
062 +
063 g → H.MS
064 DSP 4
065 f – x –
066 f P <--> S
067 RCL 5
068 RCL 6
069 g → P
070 RCL 4
071 ×
072 f – x –
073 f LBL 4
074 h R ↓
075 1
076 8
077 0
078 +
079 DSP 0
080 f – x –
081 f P <--> S
082 RCL A
083 f cos
084 RCL 5
085 f cos
086 ×
087 RCL B
088 f cos
089 ×
090 RCL 5
091 f sin
092 RCL B
093 f sin
094 ×
095 +
096 g sin⁻¹
097 h RTN
098 f LBL 3
099 RCL 6
100 RCL 4
101 ×
102 RCL C
103 +
104 RCL 7
105 –
106 STO A
107 RCL 1
108 RCL 6
109 ×
110 RCL 3
111 +
112 RCL 6
113 ×
114 RCL 2
115 +
116 RCL A
117 f sin
118 RCL 9
119 ×
120 f P <--> S
121 STO 8
122 –
123 STO 5
124 RCL 1
125 f P <--> S
126 RCL 6
127 ×
128 f P <--> S
129 RCL 3
130 +
131 f P <--> S
132 RCL 6
133 ×
134 f P <--> S
135 RCL 2
136 +
137 f P <--> S
138 RCL B
139 f cos
140 RCL 8
141 ×
142 –
143 RCL 9
144 f P <--> S
145 RCL A
146 f cos
147 ×
148 STO 7
149 RCL B
150 f sin
151 STO × 8
152 ×
153 +
154 STO 6
155 RCL 0
156 STO × 7
157 STO × 8
158 RCL D
159 STO + 7
160 RCL E
161 STO + 8
162 RCL 7
163 RCL 8
164 g → P
165 STO 9
166 RCL 5
167 RCL 7
168 ×
169 RCL 6
170 RCL 8
171 ×
172 +
173 RCL 9
174 g x²
175 ÷
176 f P <--> S
177 STO – 6
178 h RTN
```

3. For immersion: press B.

The display flashes the UT of immersion as predicted for the specified location (H.MS format, to the nearest whole second), then the Position Angle (*P*, to the nearest degree) and the programme ends by displaying the star's altitude (*h*, to the nearest whole degree; a negative value would indicate that the star is below the horizon from the specified location).

For emersion: press D.

The display flashes UT and *P*, and ends by showing *h*.

For least distance: press C.

The programme gives UT of the nearest approach to the centre of the Moon, the least distance expressed in terms of the Moon's radius (less than 1.0 indicates an occultation will take place as seen from the specified location, provided that *h* is positive; more than 1.0 shows that the star is not, in fact, occulted at the given location), then *P*, and finally, *h*.

4. To make a prediction for another location:

Enter, for the new station, φ, STO 5
λ, STO 7
$\rho \sin\varphi'$, STO 8
$\rho \cos\varphi'$, STO 9

and return to Step 3.

Test: Use the data from the search-routine example for *SAO* 094 027, α Tau, to make a prediction for its occultation on 1978, August 26, at Greenwich.

Load Programme 54/55A

Enter:			
4.344 147,	press A	(α of star)	
16.275 60,	press R/S	(δ of star)	
22.153 347 3,	R/S	(GST at 0^h UT)	
3,	R/S	(Central hour, ET)	
49,	R/S	(ΔT in seconds)	
4.320 718 6,	R/S	(α_1 at 2^h ET)	Data from lunar ephemeris
4.341 321 3,	R/S	(α_2 at 3^h ET)	
4.361 922 1,	R/S	(α_3 at 4^h ET)	
16.515 458,	R/S	(δ_1 at 2^h ET)	
16.553 311,	R/S	(δ_2 at 3^h ET)	
16.590 664,	R/S	(δ_3 at 4^h ET)	
0.545 780,	R/S	(π_1 at 2^h ET)	
0.545 646,	R/S	(π_2 at 3^h ET)	
0.545 514,	R/S	(π_3 at 4^h ET)	

When π appears:

Enter		
51.5,	STO 5	(φ at Greenwich)
0,	STO 7	(λ)
0.778 97,	STO 8	($\rho \sin\varphi'$)
0.623 80,	STO 9	($\rho \cos\varphi'$)

Now load Programme 54B

Press B: Immersion $= 1^h\ 56^m\ 21^s$ UT
P $= 28°$
h $= 28°$

Press D: Emersion = $2^h\,41^m\,15^s$ UT

P = 306°

h = 35°

Press C: Least distance = $2^h\,18^m\,28^s$ UT

= 0.754 3 r

P = 347°

h = 31°

Now, find the situation at Edinburgh:

Enter: 55.925, STO 5 (φ)

3.182 5, STO 7 (λ)

0.824 67, STO 8 (ρ sinφ')

0.561 58, STO 9 (ρ cosφ')

Press B: Immersion = $2^h\,13^m\,09^s$ UT

P = 8°

h = 28°

Press D: Emersion = $2^h\,35^m\,46^s$ UT

P = 328°

h = 31°

Press C: Least distance = $2^h\,24^m\,23^s$ UT

= 0.941 0 r

P = 348°

h = 29°

Note that at Edinburgh the least distance is 0.941 0 Moon radii, close to unity; this indicates that Edinburgh is not far from the northern limit of the occultation zone. We shall examine this point with Programme 55B.

(55)B **HP-67**

Grazing occultations.

1. Load Programme 54/55A and enter the quantities α ☾ to π_3 as indicated. As we are not predicting for a fixed point, the additional data entered at this stage for Programme 54B are not required and must not be entered.

2. Load the following programme from a magnetic card:

001	f LBL 0	047	f sin	093	g x^2	139	.
002	RCL 1	048	RCL 8	094	÷	140	0
003	RCL A	049	×	095	−	141	6
004	×	050	f P ←→ S	096	STO A	142	4
005	RCL 3	051	STO 8	097	RCL 8	143	3
006	+	052	−	098	STO × 5	144	1
007	RCL A	053	STO 5	099	RCL 7	145	1
008	×	054	f GSB 0	100	RCL 6	146	STO ÷ 9
009	RCL 2	055	RCL B	101	×	147	1
010	+	056	f cos	102	STO − 5	148	h F ? 0
011	h RTN	057	f P ←→ S	103	RCL 9	149	CHS
012	f LBL D	058	RCL 7	104	STO ÷ 5	150	RCL 5
013	h CF 0	059	×	105	RCL 4	151	−
014	GTO 1	060	−	106	STO × 5	152	RCL 9
015	f LBL E	061	RCL 8	107	RCL B	153	×
016	h SF 0	062	f P←→ S	108	f cos	154	f P ←→ S
017	f LBL 1	063	RCL E	109	.	155	STO + 6
018	STO 5	064	f cos	110	9	156	DSP 5
019	0	065	×	111	9	157	*h pause
020	STO A	066	STO 7	112	3	158	h ABS
021	STO 6	067	RCL B	113	2	159	EEX
022	9	068	f sin	114	8	160	5
023	h ST I	069	STO × 8	115	×	161	CHS
024	f LBL 2	070	×	116	f P ←→ S	162	g $x > y$
025	RCL 6	071	+	117	RCL 8	163	GTO 3
026	f tan	072	STO 6	118	×	164	f DSZ
027	RCL 9	073	RCL 0	119	RCL B	165	GTO 2
028	×	074	STO × 7	120	f sin	166	GTO 4
029	g tan^{-1}	075	STO × 8	121	RCL E	167	f LBL 3
030	f cos	076	RCL 3	122	f cos	168	RCL 6
031	STO 8	077	STO + 8	123	×	169	R/S
032	h last x	078	RCL D	124	RCL 7	170	RCL A
033	f sin	079	STO + 7	125	f P ←→ S	171	RCL 0
034	RCL 9	080	RCL 7	126	STO 6	172	+
035	×	081	RCL 8	127	×	173	DSP 4
036	STO 7	082	g → P	128	+	174	g → H.MS
037	RCL A	083	STO 9	129	RCL 7	175	R/S
038	RCL 4	084	RCL A	130	×	176	RCL E
039	×	085	RCL 5	131	RCL E	177	f cos
040	RCL C	086	RCL 7	132	f sin	178	RCL B
041	+	087	×	133	RCL 8	179	f cos
042	RCL 5	088	RCL 6	134	×	180	×
043	−	089	RCL 8	135	RCL 6	181	RCL 6
044	STO E	090	×	136	×	182	f cos
045	f GSB 0	091	+	137	+	183	×
046	RCL E	092	RCL 9	138	STO ÷ 9	184	RCL B

* Or, if you prefer, f − x −.

185	f sin	192	R/S	199	0	206	3
186	RCL 6	193	f P ←→ S	200	h F ? 0	207	6
187	f sin	194	RCL 7	201	CHS	208	0
188	×	195	RCL 8	202	−	209	+
189	+	196	g → P	203	f P ←→ S	210	h RTN
190	g $\sin^{-1}$	197	h R ↓	204	f $x > 0$		
191	DSP 2	198	9	205	h RTN		

3. Enter a longitude (usually an integer, but if not, then in decimal degrees; negative if E of Greenwich).

4. For a northern limit, press D.
   For a southern limit, press E.

5. The programme starts from $\varphi = 0°$. Successive corrections, $\Delta\varphi$ (in degrees), are briefly displayed during the pauses, until finally the limiting latitude is given, in decimal degrees. (If, after 9 iterations, $|\Delta\varphi|$ is still $> 10^{-5}$, then 'Error' appears.)
   Then press R/S to obtain the UT (H.MS format), press R/S again to obtain h (in decimal degrees), and finally press R/S to obtain P (in decimal degrees).

6. Return to Step 3 and enter another longitude (usually separated by 2°, but closer plots can be made if desired to enable an accurate limit line to be drawn on a map). Repeat Steps 4 and 5.

7. When enough limiting latitudes have been obtained, draw the limit line on a map. The prediction is that places on this line will (subject to the accuracy of the lunar ephemerides) observe a grazing occultation.

Test: In the prediction for the occultation of α Tau at Edinburgh on 1978. August 26 it was noted from Programme 54B that the least distance was 0.941 0, Thus, this station is near the northern limit of the occultation zone. Find and list the northern limits for longitudes 8°, 6°, 4° and 2° W.

Load Programme 54/55A.

Enter:	4.344 147,	press A	(α of star)	
	16.275 60,	press R/S	(δ of star)	
	22.153 347 3,	R/S	(GST at 0^h UT)	
	3,	R/S	(Central hour, ET)	
	49,	R/S	(ΔT in seconds)	
	4.320 718 6,	R/S	(α_1 at 2^h ET)	Data from lunar ephemeris
	4.341 321 3,	R/S	(α_2 at 3^h ET)	
	4.361 922 1,	R/S	(α_3 at 4^h ET)	
	16.515 458,	R/S	(δ_1 at 2^h ET)	
	16.553 311,	R/S	(δ_2 at 3^h ET)	
	16.590 664,	R/S	(δ_3 at 4^h ET)	
	0.545 780,	R/S	(π_1 at 2^h ET)	
	0.545 646,	R/S	(π_2 at 3^h ET)	
	0.545 514,	R/S	(π_3 at 4^h ET)	

When π appears, load Programme 55B.

* Enter 8, press D; note the northern limit in decimal degrees
  Press R/S and note the UT
  Press R/S for h
  Press R/S for P

Return to * and repeat for the other longitudes.

Tabulate the results:

Longitude	8° W	6° W	4° W	2° W
Latitude	55°.929 93	56°.748 91	57°.552 51	58°.337 62
	(55° 55′ 47″.7)	(56° 44′ 56″.1)	(57° 33′ 09″.0)	(58° 20′ 15″.4)
UT	$2^h 22^m 13^s$	$2^h 24^m 29^s$	$2^h 26^m 48^s$	$2^h 29^m 09^s$
h	26°.33	27°.57	28°.74	29°.84
P	348°.06	348°.12	348°.20	348°.29

Plotting the northern limit on a map of the British Isles gives the same indication as that for Occultation No. 6 on the outline map on p 25 of the 1978 *Handbook* of the BAA. The line passes just to the south of the Hebridean islands of Tiree and Coll, cuts the mainland of Scotland just south of Ardnamurchan Point, crosses northwest across Lochaber and Inverness districts of the Highland region to Inverness itself, and finally cuts the east coast west of Nairn.

For a southern limit, enter 30, CHS (for 30°E) and press E. The latitude given is 25°.386 27 (25° 23′ 10″.6). At 20°E, the southern limit is 21°.138 70 (21° 08′ 19″.3). The southern limit of this occultation thus crosses Egypt, near Al-Kharijah (El Kharga), with the star fairly high in the sky.

To make a search for possible occultations of stars by the Moon.

1. Extract from the ephemeris:
   apparent α for the moon
   apparent δ
   π (horizontal parallax)

at the start of the period under review, and for the subsequent integral hours of ET. The object of the exercise is to identify the stars with which the Moon is about to come into conjunction, and to eliminate those which do not lie within the declination limits where an occultation is possible.

2. From the appropriate declination band of the *SAO Star Catalogue* (in which lies $\delta☾$ at the commencement of the period) extract *SAO* Number, magnitude, 1950.0 α^*, μ_α, δ^*, μ_δ, for those stars within the limits $\alpha☾ - 2^m$ (to allow for the effect of precession on the 1950.0 coordinates) and $\alpha☾ +$ the number of hours to be covered by the review (normally 2 to 3 hours), and $\pm 3°$ of $\delta☾$ at the start of the period. Exclude stars fainter than $8^m.0$.

3. Tabulate the preliminary data:

1	2	3	4	5	6	7	8
SAO Number	Mag.	Hour	α^*	δ^*	$\delta\delta☾$	$A - \delta^*$	$B - \delta^*$

For the first entry, complete Columns 3, 4 and 5 for the Moon at the integral hour starting the period, then list the stars, completing Columns 1, 2, 4 and 5. Intersperse, at the appropriate α, the data for the Moon at successive integral hours.

4. For each hourly section between consecutive entries for the Moon, let the ET at the start of the period be T_1 and the next integral hour of ET be T_2.

5. For each such period, deduct δ at T_1 from δ at T_2 and call this quantity $\delta\delta$. Complete Column 6.

6. Compute A and B for each hourly period, from:

$$A = \delta + \delta\delta \pm (Z + \pi)$$
$$B = \delta \pm (Z + \pi)$$

where the values for δ and π are those for the start of the hourly period, $Z = 1° 40'$, and the sign for $(Z + \pi)$ agrees with the sign of $\delta\delta$ (i.e., if $\delta\delta$ is positive, *add* $(Z + \pi)$ for A, and *subtract* $(Z + \pi)$ for B; if $\delta\delta$ is negative, reverse this procedure). Complete Columns 7 and 8.

7. Mark with an obelisk (†) those stars for which the entries in Columns 7 and 8 bear the same sign. These stars are excluded from the subsequent calculations.The remainder forms a crude list of stars which might be occulted by the Moon during the review period. This preliminary selection must now be refined.

8. First, by Programme 11, reduce the 1950.0 coordinates of the retained stars to their mean places at the nearest start of a Besselian solar year. Then find each star's apparent place to the first order, by Programme 56, for the integral hour of ET immediately preceding the star's position in the crude list. The programme will interpolate the Besselian Day Numbers and compute the desired apparent place.

9. Repeat Steps 3 and 6, this time using $Z = 21' 24''$, and re-tabulate. Again, mark for exclusion any stars where the entries in Columns 7 and 8 bear the same sign.

10. The final list now contains the stars which will probably be occulted during the review period, and which can be observed with telescopes of moderate to large aperture (say, 200 mm and larger). If a smaller instrument is to be used, apply the same exclusion principles as outlined on pp 279–280 of the *Explanatory Supplement to the AE.*

Even if an occultation occurs, it may not be visible at the observer's location; the calculator programmes will confirm those which are, and the northern or southern limits for any grazing occultations.

Example. Are any stars brighter than $8^m.0$ likely to be occulted by the Moon for European observers from 1^h ET onwards on the morning of 1978, August 26?

1.

	α☾	δ☾	π☾
1^h ET	$4^h 30^m 01^s.141$	$+16° 48' 11''.07$	$54' 59''.15$
2^h	$4^h 32^m 07^s.186$	$16° 51' 54''.58$	$54' 57''.80$
3^h	$4^h 34^m 13^s.213$	$16° 55' 33''.11$	$54' 56''.46$
4^h	$4^h 36^m 19^s.221$	$16° 59' 06''.64$	$54' 55''.14$

2.

SAO	Mag.	1950.0 α*	μ_α sec	1950.0 δ*	μ_δ
093 983	6.6	$4^h 28^m 16^s.719$	+0.003 1	$+14° 59' 56''.13$	$-0''.045$
93	6.0	$4^h 29^m 00^s.180$	+0.007 0	$15° 44' 45''.45$	$-0''.026$
98	7.8	$4^h 30^m 05^s.364$	+0.001 2	$19° 14' 38''.12$	$-0''.032$
094 002	6.2	$4^h 30^m 39^s.055$	+0.000 8	$17° 54' 46''.03$	$-0''.023$
04	6.5	$4^h 30^m 46^s.207$	+0.001 1	$16° 13' 09''.85$	$-0''.023$
07	4.7	$4^h 31^m 00^s.440$	+0.006 9	$14° 44' 27''.45$	$-0''.025$
15	8.0	$4^h 31^m 48^s.401$	+0.003 6	$17° 38' 44''.97$	$-0''.045$
18	7.4	$4^h 32^m 05^s.966$	−0.000 2	$16° 53' 35''.65$	$-0''.076$
19	7.1	$4^h 32^m 09^s.458$	−0.000 8	$17° 05' 56''.12$	$+0''.014$

20	8.0	$4^h 32^m 42^s.049$	+0.001 3	16° 02′ 35″.72	−0″.038
22	6.6	$4^h 32^m 46^s.532$	0	19° 46′ 48″.80	−0″.016
27	1.1	$4^h 33^m 02^s.896$	+0.004 5	16° 24′ 37″.51	−0″.189
31	7.8	$4^h 33^m 38^s.674$	−0.000 7	19° 39′ 33″.13	−0″.004
33	6.7	$4^h 33^m 48^s.998$	+0.006 8	15° 46′ 08″.22	−0″.028
34	7.6	$4^h 34^m 04^s.170$	+0.002 6	15° 09′ 40″.35	−0″.037
36	7.2	$4^h 34^m 20^s.154$	−0.001 2	18° 26′ 35″.17	−0″.005
40	7.8	$4^h 34^m 41^s.048$	+0.006 5	15° 02′ 48″.66	−0″.041
43	5.8	$4^h 35^m 17^s.491$	+0.006 4	15° 56′ 05″.25	−0″.022
51	5.1	$4^h 36^m 17^s.709$	+0.002 7	15° 42′ 10″.83	−0″.071

3. $A_1 = 16° 48' 11''.07 + 3' 43''.51 + (1° 40' + 54' 59''.15)$
   $= 19° 26' 53''.73\ (19°.448\ 259)$
   $B_1 = 16° 48' 11''.07 - (1° 40' + 54' 59''.15)$
   $= 14° 13' 11''.92\ (14°.219\ 978)$
   $A_2 = 16° 51' 54''.58 + 3' 38''.53 + (1° 40' + 54' 57''.80)$
   $= 19° 30' 30''.91\ (19°.508\ 586)$
   $B_2 = 16° 51' 54''.58 - (1° 40' + 54' 57''.80)$
   $= 14° 16' 56''.78\ (14°.282\ 439)$
   $A_3 = 16° 55' 33''.11 + 3' 33''.53 + (1° 40' + 54' 56''.46)$
   $= 19° 34' 03''.10\ (19°.567\ 528)$
   $B_3 = 16° 55' 33''.11 - (1° 40' + 54' 56''.46)$
   $= 14° 20' 36''.65\ (14°.343\ 514)$

4. Crude list:

SAO	Mag.	Time	α^* h m s	δ^* ° ′ ″	$\delta\delta$ ′ ″	$A-\delta^*$ ° ′ ″	$B-\delta^*$ ° ′ ″
Moon		1^h ET	4 30 01.141	+16 48 11.07	+3 43.51		
093 983	6.6		4 28 16.719	14 59 56.13		+4 26 57.6	−0 46 44.2
93	6.0		4 29 00.180	15 44 45.45		+3 42 08.3	−1 31 33.5
98	7.8		4 30 05.364	19 14 38.12		+0 12 15.6	−5 01 26.2
094 002	6.2		4 30 39.055	17 54 46.03		+1 32 07.7	−3 41 34.1
04	6.5		4 30 46.207	16 13 09.85		+3 13 43.9	−1 59 57.9
07	4.7		4 31 00.440	14 44 27.45		+4 42 26.3	−0 31 15.5
15	8.0		4 31 48.401	17 38 44.97		+1 48 08.8	−3 25 33.0
18	7.4		4 32 05.966	16 53 35.65		+2 33 18.1	−2 40 23.7
Moon		2^h ET	4 32 07.186	16 51 54.58	+3 38.53		
19	7.1		4 32 09.458	17 05 56.12		+2 24 34.8	−2 48 59.3
20	8.0		4 32 42.049	16 02 35.72		+3 27 55.2	−1 45 38.9
22†	6.6		4 32 46.532	19 46 48.80		−0 16 17.9	−5 29 52.0
27	1.1		4 33 02.896	16 24 37.51		+3 05 53.4	−2 07 40.7
31†	7.8		4 33 38.674	19 39 33.13		−0 09 02.2	−5 22 36.3
33	6.7		4 33 48.998	15 46 08.22		+3 44 22.7	−1 29 11.4
34	7.6		4 34 04.170	15 09 40.35		+4 20 50.6	−0 52 43.6
Moon		3^h ET	4 34 13.213	16 55 33.11	+3 33.53		
36	7.2		4 34 20.154	18 26 35.17		+1 07 27.9	−4 05 58.5
40	7.8		4 34 41.048	15 02 48.66		+4 31 14.4	−0 42 12.0
43	5.8		4 35 17.491	15 56 05.25		+3 37 57.8	−1 35 28.6
51	5.1		4 36 17.709	15 42 10.83		+3 51 52.3	−1 21 34.2
Moon		4^h ET	4 36 19.221	16 59 06.64			

† *SAO* 094 022 and 031 are excluded (similarity of sign in Columns 7 and 8.)

5. Mean places of retained stars at 1979.0;

SAO	α^*	δ^*	μ_α sec	$\mu\delta$
093 983	$4^h\,29^m\,55^s.54$	$+15°\,03'\,39''.3$	+0.003 1	$-0''.045$*
93	$4^h\,30^m\,39^s.63$	$15°\,48'\,27''.5$	+0.007 0	$-0''.026$
98	$4^h\,31^m\,47^s.07$	$19°\,18'\,17''.4$	+0.001 2	$-0''.032$
094 002	$4^h\,32^m\,19^s.83$	$17°\,58'\,24''.3$	+0.000 8	$-0''.023$
04	$4^h\,32^m\,25^s.83$	$16°\,16'\,47''.8$	+0.001 1	$-0''.023$
07	$4^h\,32^m\,39^s.24$	$14°\,48'\,04''.8$	+0.006 9	$-0''.025$
15	$4^h\,33^m\,29^s.10$	$17°\,42'\,19''.9$	+0.003 6	$-0''.045$
18	$4^h\,33^m\,46^s.04$	$16°\,57'\,09''.0$	−0.000 2	$-0''.076$
19	$4^h\,33^m\,49^s.66$	$17°\,09'\,31''.9$	−0.000 8	$+0''.014$
20	$4^h\,34^m\,21^s.60$	$16°\,06'\,08''.7$	+0.001 3	$-0''.038$
27	$4^h\,34^m\,42^s.79$	$16°\,28'\,05''.3$	+0.004 5	$-0''.189$
33	$4^h\,35^m\,28^s.54$	$15°\,49'\,38''.9$	+0.006 8	$-0''.028$
34	$4^h\,35^m\,43^s.18$	$15°\,13'\,10''.2$	+0.002 6	$-0''.037$
36	$4^h\,36^m\,01^s.32$	$18°\,30'\,05''.3$	−0.001 2	$-0''.005$
40	$4^h\,36^m\,20^s.10$	$15°\,06'\,16''.9$	+0.006 5	$-0''.041$
43	$4^h\,36^m\,57^s.16$	$15°\,59'\,32''.6$	+0.006 4	$-0''.022$
51	$4^h\,37^m\,57^s.13$	$15°\,45'\,34''.4$	+0.002 7	$-0''.071$

6. Besselian Day Numbers:

	A	*B*	*C*	*D*	*E*	τ
Aug 26	$-6''.914$	$+9''.023$	$+16''.669$	$-9''.506$	0^s	−0.350 3
Aug 27	$-6''.858$	$+9''.022$	$+16''.815$	$-9''.199$	0^s	−0.347 6

$m = 46''.107\,0$ $n = 20''.040\,1$ $\varepsilon = 23°.442\,059$

These values are input for Programme 56 for interpolation to the integral hour of ET during the programme run.

7. Apparent places of retained stars, 1978, August 26:

SAO	α	δ	
093 983	$4^h\,29^m\,54^s.26$	$+15°\,03'\,30''.4$	At 1^h ET
93	$4^h\,30^m\,38^s.34$	$15°\,48'\,18''.3$	
98	$4^h\,31^m\,45^s.76$	$19°\,18'\,07''.0$	
094 002	$4^h\,32^m\,18^s.52$	$17°\,58'\,14''.4$	
04	$4^h\,32^m\,24^s.53$	$16°\,16'\,38''.5$	
07	$4^h\,32^m\,37^s.94$	$14°\,47'\,56''.0$	
15	$4^h\,33^m\,27^s.79$	$17°\,42'\,10''.1$	
18	$4^h\,33^m\,44^s.73$	$16°\,56'\,59''.5$	
19	$4^h\,33^m\,48^s.35$	$17°\,09'\,22''.3$	At 2^h ET
20	$4^h\,34^m\,20^s.29$	$16°\,05'\,59''.5$	
27	$4^h\,34^m\,41^s.47$	$16°\,27'\,56''.0$	
33	$4^h\,35^m\,27^s.22$	$15°\,49'\,29''.8$	
34	$4^h\,35^m\,41^s.86$	$15°\,13'\,01''.3$	
36	$4^h\,35^m\,59^s.99$	$18°\,29'\,55''.2$	At 3^h ET
40	$4^h\,36^m\,18^s.78$	$15°\,06'\,08''.1$	
43	$4^h\,36^m\,55^s.83$	$15°\,59'\,23''.5$	
51	$4^h\,37^m\,55^s.80$	$15°\,45'\,25''.4$	

* The proper options are found to be unchanged from the 1950.0 values.

8. Re-compute A and B as in Step 3, now with $Z = 21' 24''$:

$A_1 = 18° 08' 17''.72$
$B_1 = 15° 31' 47''.92$
$A_2 = 18° 11' 54''.91$
$B_2 = 15° 35' 32''.77$
$A_3 = 18° 15' 27''.10$
$B_3 = 15° 39' 12''.64$
$A_4 = 18° 18' 54''.31$
$B_4 = 15° 42' 47''.50$

9. Refine the conjunctions:

SAO	Mag.	Time	α^*	δ^*	$\delta\delta$	$A - \delta^*$	$B - \delta^*$
			h m s	° ′ ″	′ ″	° ′ ″	° ′ ″
Moon		1ʰ ET	4 30 01.141	+16 48 11.07	+3 43.51		
093 983†	6.6		4 29 54.26	15 03 30.4		+3 04 47.3	+0 28 17.5
93	6.0		4 30 38.34	15 48 18.3		+2 19 59.4	−0 16 30.4
98†	7.8		4 31 45.76	19 18 07.0		−0 09 49.3	−3 46 19.1
Moon		2ʰ ET	4 32 07.186	16 51 54.58	+3 38.53		
094 002	6.2		4 32 18.52	17 58 14.4		+0 13 40.5	−2 22 41.6
04	6.5		4 32 24.53	16 16 38.5		+1 55 16.4	−0 41 05.7
07†	4.7		4 32 37.94	14 47 56.0		+3 23 58.9	+0 47 36.8
15	8.0		4 33 27.79	17 42 10.1		+0 29 44.8	−2 06 37.3
18	7.4		4 33 44.73	16 56 59.5		+1 14 55.4	−1 21 26.7
19	7.1		4 33 48.35	17 09 22.3		+1 02 32.6	−1 33 49.5
Moon		3ʰ ET	4 34 13.213	16 55 33.11	+3 33.53		
20	8.0		4 34 20.29	16 05 59.5		+2 09 27.6	−0 26 46.8
27	1.1		4 34 41.47	16 27 56.0		+1 47 31.1	−0 48 43.3
33	6.7		4 35 27.22	15 49 29.8		+2 25 57.3	−0 10 17.1
34†	7.6		4 35 41.86	15 13 01.3		+3 02 25.8	+0 26 11.4
36†	7.2		4 35 59.99	18 29 55.2		−0 14 28.1	−2 50 42.5
40†	7.8		4 36 18.78	15 06 08.1		+3 09 19.0	+0 33 04.6
Moon		4ʰ ET	4 36 19.221	16 59 06.64	+3 28.54		
43	5.8		4 36 55.83	15 59 23.5		+2 19 30.8	−0 16 36.0
51	5.1		4 37 55.80	15 45 2.54		+2 33 28.9	−0 02 37.9

† *SAO* 093 983, 098, 094 007, 034, 036 and 040 are excluded for similarity of sign in Columns 7 and 8.

10. The final list of stars likely to be occulted for European observers on the morning of 1978, August 26 is:

SAO	Mag.	α^*	δ^*
093 993	6.0	4ʰ 30ᵐ 38ˢ.34	+15° 48′ 18″.3
094 002	6.2	4ʰ 32ᵐ 18ˢ.52	17° 58′ 14″.4
04	6.5	4ʰ 32ᵐ 24ˢ.53	16° 16′ 38″.5
15	8.0	4ʰ 33ᵐ 27ˢ.79	17° 42′ 10″.1
18	7.4	4ʰ 33ᵐ 44ˢ.73	16° 56′ 59″.5
19	7.1	4ʰ 33ᵐ 48ˢ.35	17° 09′ 22″.3
20	8.0	4ʰ 34ᵐ 20ˢ.29	16° 05′ 59″.5
27	1.1	4ʰ 34ᵐ 41ˢ.47	16° 27′ 56″.0
33	6.7	4ʰ 35ᵐ 27ˢ.22	15° 49′ 29″.8

43	5.8	$4^h\,36^m\,55^s.83$	15° 59′ 23″.5
51	5.1	$4^h\,37^m.55^s.80$	15° 45′ 25″.4

11. Now proceed to Programmes 54/55A, 54B and 55B. These programmes will confirm whether occultations will occur at the selected location and, for any grazing occultations, the northern (and/or southern) limits. As an example, we took the case of the brightest star in the list, which is α Tau, and used the foregoing data in both the rigorous programmes.

If, for practice, you try other stars in the list of 'possibles' you will find some interesting results; for example, *SAO* 094 033 is not, in fact, occulted (the least distance at Greenwich is 1.888 4 at $2^h\,23^m\,44^s$ UT); neither is *SAO* 094 051, which is 4′ farther south. But *SAO* 094 020, which lies a little farther north in declination, is occulted at $1^h\,30^m\,21^s$ UT (at Greenwich). This star also has an interesting southern track for a grazing occultation, which you can plot by running Programme 55B. The track crosses southern Spain and cuts the east coast a few miles north of Castellón de la Plana. In southwest Spain the star will be at low altitude, but it will be 25° above the horizon on the east coast.

Some of the other stars in the list are occulted, but not as seen from Europe. *SAO* 094 043 is visible from Iceland, but the southern limit of this occultation misses the most northly part of Norway and Sweden.

(56) **HP-67**

Reduction of the mean place of a star at the nearest start of a Besselian solar year to the apparent place at any integral hour, to the first order.

---

1. Load the programme from a magnetic card:

001	f LBL A	051	RCL 8	101	RCL 0	151	f sin
002	1	052	–	102	f cos	152	CHS
003	9	053	h last x	103	RCL 2	153	RCL B
004	h ST I	054	h $x \longleftrightarrow y$	104	f cos	154	×
005	h R ↓	055	RCL 6	105	h $1/x$	155	STO + 6
006	STO (i)	056	×	106	×	156	RCL 5
007	f DSZ	057	+	107	RCL C	157	RCL 2
008	DSP 0	058	h ST I	108	×	158	f cos
009	f GSB 1	059	h π	109	1	159	×
010	RCL 6	060	DSP 2	110	5	160	RCL 0
011	RCL 5	061	h RTN	111	÷	161	f sin
012	÷	062	f LBL C	112	STO + 7	162	RCL 2
013	STO 4	063	f H ←	113	RCL 0	163	f sin
014	RCL 7	064	1	114	f sin	164	×
015	f tan	065	5	115	RCL 2	165	–
016	STO 5	066	×	116	f cos	166	RCL C
017	DSP 2	067	STO 0	117	h $1/x$	167	×
018	h π	068	DSP 3	118	×	168	STO + 6
019	h RTN	069	R/S	119	RCL D	169	RCL 0
020	f LBL B	070	STO 1	120	×	170	f cos
021	2	071	R/S	121	1	171	RCL 2
022	4	072	f H ←	122	5	172	f sin
023	÷	073	STO 2	123	÷	173	×
024	CHS	074	R/S	124	STO + 7	174	RCL D
025	1	075	STO 3	125	RCL E	175	×
026	+	076	f LBL D	126	STO + 7	176	STO + 6
027	STO 6	077	RCL 4	127	RCL 1	177	RCL 3
028	f P ←→ S	078	RCL 0	128	h RC I	178	h RC I
029	RCL 9	079	f sin	129	×	179	×
030	RCL 8	080	RCL 2	130	STO + 7	180	STO + 6
031	f GSB 2	081	f tan	131	RCL 7	181	RCL 6
032	STO A	082	×	132	3	182	3
033	RCL 7	083	+	133	6	183	6
034	RCL 6	084	RCL A	134	0	184	0
035	f GSB 2	085	×	135	0	185	0
036	STO B	086	1	136	÷	186	÷
037	RCL 5	087	5	137	RCL 0	187	RCL 2
038	RCL 4	088	÷	138	1	188	+
039	f GSB 2	089	STO 7	139	5	189	g → H.MS
040	STO C	090	RCL 0	140	÷	190	DSP 5
041	RCL 3	091	f cos	141	+	191	h RTN
042	RCL 2	092	RCL 2	142	g → H.MS	192	f LBL 1
043	f GSB 2	093	f tan	143	DSP 6	193	R/S
044	STO D	094	×	144	f – x –	194	STO (i)
045	RCL 1	095	RCL B	145	RCL 0	195	f DSZ
046	RCL 0	096	×	146	f cos	196	4
047	f GSB 2	097	1	147	RCL A	197	h RC I
048	STO E	098	5	148	×	198	g $x = y$
049	f P ←→ S	099	÷	149	STO 6	199	h RTN
050	RCL 9	100	STO + 7	150	RCL 0	200	GTO 1

201	f LBL 2	204	h $x \longleftrightarrow y$	207	f P $\longleftrightarrow$ S	210	h RTN
202	–	205	f P $\longleftrightarrow$ S	208	×		
203	h last x	206	RCL 6	209	+		

For the start of the selected day, let the Besselian Day Numbers bear the suffix 1 (e.g., A_1, B_1, etc.). Let the numbers for the start of the next day bear the suffix 2. Ignore J and J' which are only required for reductions to the second order.

2. Enter A_1 and press A.

Then, at each succeeding halt, enter in turn:

A_2, B_1, B_2, C_1, C_2, D_1, D_2, E_1, E_2, τ_1, τ_2, ϵ (in decimal degrees), m (seconds of arc) and n (seconds of arc). ϵ need only be taken for the start of the selected day; m and n for the nearest beginning of a year.

After n is entered, π (3.14) will be displayed to signify completed data entry for interpolation.

3. Enter the integral hour of ET for which the apparent places are required, and press B.

The Day Numbers are interpolated for the required hour and stored ready for further use. (The original two sets of Day Numbers are also retained, and can be used for re-interpolation to any other integral hour of ET on the same day; this is extremely useful for occultation work.)

π is again displayed to show that interpolation has been completed.

4. Enter the mean place of the star at the beginning of the nearest Besselian solar year:

α (H.MS), press C
μ_α (seconds of time per annum), press R/S
δ (D.MS), press R/S
μ_δ (seconds of arc), press R/S.

5. Reduction to the apparent place at the required time is now carried out. The display pauses to flash the apparent RA (H.MS format) at the integral hour of ET selected, and the programme concludes by displaying the apparent dec.

6. If the apparent place of the same star is required for a different integral hour of ET on the same day, enter the integer for the hour, press B; when π appears, press D.

7. For the next star, return to Step 4.

8. If $\Delta\alpha$ is required, press RCL 7.
If $\Delta\delta$ is required, press RCL 6.

Test: Find the apparent place of *SAO* 094 051 at 3^h ET on 1978, August 26, given the mean place at 1979.0 is:

$\alpha = 4^h\,37^m\,57^s.13$, $\delta = +15°\,45'\,34''.4$, $\mu_\alpha = +0^s.002\,7$, $\mu_\delta = -0''.071$

and the Besselian Day Numbers are:

	A	B	C	D	E	τ
Aug 26	$-6''.914$	$+9''.023$	$+16''.669$	$-9''.506$	0	$-0.350\,3$
Aug 27	$-6''.858$	$+9''.022$	$+16''.815$	$-9''.199$	0	$-0.347\,6$

At the start of the day, $\epsilon = 23°.442\,059$; for 1979.0 $m = 46''.107\,1$, $n = 20''.040\,1$. These values may be used throughout the day.

The result given by the programme is:

$\alpha = 4^h\,37^m\,55^s.80$, $\delta = +15°\,45'\,25''.4$

These are the coordinates used in the example of an occultation search.

---

(57) **HP-67**

To compute the Besselian elements of an occultation.

1. Load the programme from a magnetic card:

001	f LBL A	043	RCL 0	085	÷	127	RCL 1
002	DSP 4	044	RCL 6	086	×	128	f P ←→ S
003	f H ←	045	–	087	f P ←→ S	129	.
004	STO 0	046	RCL E	088	STO 0	130	0
005	R/S	047	×	089	f P ←→ S	131	0
006	f H ←	048	STO 0	090	RCL C	132	0
007	STO 1	049	RCL 2	091	RCL 2	133	0
008	R/S	050	RCL 6	092	RCL 5	134	3
009	f H ←	051	–	093	÷	135	6
010	STO 2	052	RCL E	094	×	136	×
011	R/S	053	×	095	f P ←→ S	137	RCL 2
012	f H ←	054	STO 2	096	STO 1	138	×
013	STO 3	055	RCL 7	097	RCL 0	139	RCL D
014	R/S	056	f sin	098	–	140	×
015	f H ←	057	STO D	099	STO 2	141	+
016	3	058	RCL 1	100	f P ←→ S	142	f P ←→ S
017	6	059	f cos	101	RCL 1	143	STO 4
018	0	060	1	102	RCL 4	144	RCL 3
019	0	061	5	103	÷	145	–
020	STO E	062	×	104	f P ←→ S	146	STO 5
021	×	063	STO B	105	RCL 0	147	RCL 3
022	STO 4	064	RCL 3	106	f P ←→ S	148	RCL 0
023	R/S	065	f cos	107	.	149	RCL 5
024	f H ←	066	1	108	0	150	RCL 2
025	RCL E	067	5	109	0	151	÷
026	×	068	×	110	0	152	×
027	STO 5	069	STO C	111	0	153	–
028	R/S	070	RCL 1	112	3	154	STO 6
029	f H ←	071	RCL 7	113	6	155	f P ←→ S
030	STO 6	072	–	114	×	156	RCL 8
031	R/S	073	RCL E	115	RCL 0	157	f P ←→ S
032	f H ←	074	×	116	×	158	RCL 0
033	STO 7	075	STO 1	117	RCL D	159	RCL 2
034	R/S	076	RCL 3	118	×	160	÷
035	STO 8	077	RCL 7	119	+	161	–
036	R/S	078	–	120	f P ←→ S	162	STO 7
037	RCL E	079	RCL E	121	STO 3	163	f P ←→ S
038	÷	080	×	122	f P ←→ S	164	RCL 9
039	STO 9	081	STO 3	123	RCL 3	165	f P ←→ S
040	R/S	082	RCL B	124	RCL 5	166	–
041	f H ←	083	RCL 0	125	÷	167	STO 8
042	STO A	084	RCL 4	126	f P ←→ S	168	RCL A

169	1	179	+	189	f – x –	199	f P ←→ S
170	.	180	f P ←→ S	190	RCL 9	200	RCL 6
171	0	181	RCL 6	191	g → H.MS	201	g → H.MS
172	0	182	–	192	f – x –	202	DSP 6
173	2	183	RCL 9	193	RCL 6	203	f – x –
174	7	184	–	194	f – x –	204	RCL 7
175	3	185	f P ←→ S	195	RCL 2	205	DSP 5
176	8	186	STO 9	196	f – x –	206	g → H.MS
177	RCL 7	187	RCL 8	197	RCL 5	207	h RTN
178	×	188	g → H.MS	198	f – x –		

2. Enter the following data:
   α at T_1 (H.MS format); press A (apparent RA of Moon at integral hour of ET immediately preceding the conjunction with the star)
   δ at T_1 (D.MS); press R/S (apparent dec. of Moon)
   α at T_2 (H.MS); press R/S (apparent RA of Moon at integral hour of ET immediately following the conjunction; T_1 and T_2 are consecutive hours)
   δ at T_2 (D.MS); R/S
   π at T_1 (D.MS); R/S (horizontal parallax)
   π at T_2 (D.MS); R/S
   α^* (H.MS); R/S (apparent RA of star at T_1)
   δ^* (D.MS); R/S (apparent dec. of star at T_1)
   T_1 (an integer); R/S (hour of ET preceding conjunction)
   ΔT (in seconds); R/S (ET – UT)
   Apparent GST at 0^h UT (H.MS) (= EST at 0^h ET)
   Press R/S

3. The programme pauses to flash, in turn:
   T_0 (H.MS) (UT of conjunction in RA)
   H_0 (H.MS) (Greenwich hour angle of the star at T_0)
   Y (y at T_0)
   x', y' } the hourly variations of x and y
   α^* (H.MS)
   and ends by displaying δ^*.

Test: Find the Besselian elements of the occultation of α Tau on 1978, August 26, given the following data:

$\alpha\ T_1 = 4^h\ 34^m\ 13^s.213$
$\delta\ T_1 = +16°\ 55'\ 33''.11$
$\alpha\ T_2 = 4^h\ 36^m\ 19^s.221$
$\delta\ T_2 = +16°\ 59'\ 06''.64$
$\pi\ T_1 = 0°\ 54'\ 56''.46$
$\pi\ T_2 = 0°\ 54'\ 55''.14$
} from lunar ephemeris

$\alpha^* = 4^h\ 34^m\ 41^s.47$
$\delta^* = +16°\ 27'\ 56''.0$
$T_1 = 3^h$ ET
$\Delta T = 49^s$
Apparent GST at 0^h UT = $22^h\ 15^m\ 33^s.473$.

The computed elements are:	The NAO published:
$T_0 = 3^h\, 12^m\, 38^s$ ($3h\, 12^m.6$)	$3^h\, 12^m.7$
$H_0 = 20^h\, 54^m\, 02^s$ ($20^h\, 54^m.0$)	$20^h\, 54^m.1$
$Y = +0.517\,4$	$+0.517\,4$
$x' = +0.548\,6$	$+0.548\,6$
$y' = +0.065\,4$	$+0.065\,4$
$\alpha^* = 4^h\, 34^m\, 41^s.47$	$4^h\, 34^m\, 41^s.48$
$\delta^* = +16^\circ\, 27'\, 56''.0$	$+16^\circ\, 27'\, 56''.0$

There are very slight differences between the two sets of elements. This is due to the fact that when computing the apparent place of the star the required time was taken as 2^h ET, but later in the refining process it became apparent that the conjunction would actually occur between 3^h and 4^h ET, for which the Besselian elements were computed. If, now, the apparent place of the star is computed for 3^h ET, it is found to be $4^h\, 34^m\, 41^s.48$ (the declination remains unchanged), thus agreeing with the NAO value. If the elements are re-computed with this fresh value for α^* all the new elements so found agree exactly with the NAO figures.

Index